Michele Martelli
Marine Propulsion Simulation

Michele Martelli

Marine Propulsion Simulation

Managing Editor: Elisa Capello

Language Editor: Mary Boyd

DE
G

DE GRUYTER
OPEN

Published by De Gruyter Open Ltd, Warsaw/Berlin
Part of Walter de Gruyter GmbH, Berlin/Munich/Boston

ISBN 978-3-11-040149-3
e-ISBN 978-3-11-040150-9

Bibliographic information published by the Deutsche Nationalbibliothek
The Deutsche Nationalbibliothek lists this publication in the Deutsche Nationalbibliografie; detailed bibliographic data are available in the Internet at http://dnb.dnb.de.

Managing Editor: Elisa Capello
Language Editor: Mary Boyd

www.degruyteropen.com

Cover illustration: © Thinkstock/dcookd

Contents

'Models and simulations have always been an essential part of the human experience. From we get up in the morning, we build up our mental model of the world around us. Then we run a few simulations, in our mind, on how we are going to deal with the problems and people are will meet during the day. We try different approaches, evaluate the likely outcomes, and start the day with a plan. It will not protect us from failures and surprises, but it will have prepared us to deal more effectively with whatever tasks await us.'

H. Bossel

1 Introduction

This work is the result of three years of research carried out between January 2010 and December 2012 as part of The XXV cycle Ph.D course in Naval Architecture and Marine Engineering. This course is provided by the "Ph.D. School in Science and Technologies for Engineering" of the Università Degli Studi Di Genova (UNIGE).

This work belongs to the Italian academic sector ING-IND/02 (Marine Construction & Marine Systems).

Funding has been provided by a scholarship from the Regione Liguria.

The challenge of this work is to develop a multi-physics simulation platform able to represent the dynamic behaviour of a ship in the time domain. With respect to previous work available in literature, where the systems are often modeled one at a time, the aim is to merge into a unique platform three ship macro-systems that contribute to the global ship dynamics: the ship manoeuvrability, the ship propulsion plant, and the control system. In this way, it is possible to catch the mutual interaction between all the involved elements treating the ship as a whole.

Design and optimization of the propulsion plant is a crucial task of the ship design due to the fact that the global behaviour of a ship is greatly influenced by the dynamic performance of the propulsion system. Different operational speeds, acceleration, deceleration, crash stop, and heavy turning are some examples of transient situations that a propulsion system has to sustain without reducing ship safety and reliability. These aspects become crucial issues if the ship is a naval vessel. The dynamic behaviour of the propulsion system is mainly affected by the control system performance, i.e., the capacity of the control system to properly use the power necessary to perform the required task within the boundary conditions imposed by machinery or environmental constraints.

The main focus of this work is to develop and use the platform for the designs of the propulsion plant and the propulsion control system for a naval vessel in the early design stage. Using this platform, it is possible to develop the control system; to try new control logics; to choose the main engines; to test the control system under different operational conditions before the ship is built. In this way, the global design time could be reduced and the final product could be better compared to the standard design technique. It also can reduce the time and cost of the sea trial for final tuning.

To reach this goal, the propulsion control logics have been first developed in a virtual environment and then in a real PLC to increase result reliability. The PLCs, linked to the ship virtual model, has been thoroughly tested and optimized. This involves a set of technical problems. The crucial task is to develop a ship simulator able to run in real time, so particular attention has been dedicated to the choice and the development of system physical models and their computation time.

The ship motion model taking into account all six degrees of freedom has been developed, and the model for rudders and propellers has been developed considering the motion three-dimensionally. The propulsion plant model includes both the main engines and the transmission line dynamics. The propulsion control system, as it is the lead objective of this work, is a subpart of the control system that has also been modeled.

Finally, an experimental campaign has been performed to validate the simulation platform.

2 Ship Dynamics

2.1 Simulation Approach

Hundreds of years ago, buildings, boats, and machineries were first tested as prototypes before being constructed on a large scale. In naval field, the pioneer William Froude established, in 1861, a methodology by which the results of small-scale towing tank tests could be used to predict the behaviour of full-sized hulls.

With the emergence of electronics came the possibility to represent complex dynamic systems by equivalent systems of electronic components using analogies between electronic circuits (capacitors, resistors, and inductors) and mathematical operations (integration, multiplication, and derivation). With advances in computer science, the behaviour of complex systems can now be simulated on a computer without even constructing and testing the real system. The computer can quickly and precisely process any mathematical or logical formulation; these new developments have led to unimaginable possibilities in all engineering domains and scientific fields.

The system dynamic behaviour can be studied by two different approaches [17].

The first approach is to describe the system behaviour from observations of one or several identical systems, observing how they change (output) under different conditions (input). Then, the description is used to create mathematical relationships to relate input to output. This method imitates the behaviour of the real system using these relationships, which usually have nothing to do with real processes in the system. The system is not described in all of its details and functionality; it is treated as a black box.

The second approach, which is employed in this thesis, tries to physically explain the behaviour of the real system by mathematical modelling. In this case, some questions must be answered. Which and how many elements are present? How are they linked? How do they influence each other? Furthermore, a huge quantity of data and information about the elements must be available, and often this is not the case.

Using this information, the dynamic system behaviour can be simulated even for conditions not observed in the past. The system is treated as a glass box, where all of its elements and processes are schematized, and where the mutual interactions of the elements are taken into account in a holistic vision of the problem. Past and future system observations come into play at a later stage known as validation. Validation is the key by which it is determined if the system has been correctly schematized.

Simulation techniques join system modelling and computer science in order to obtain, in a virtual environment, the system behaviour over time.

The simulation models are usually composed of a set of data, Boolean logic, tables, and algebraic and differential equations, all linked to each other.

2.2 Literature Review

Different works (articles, Ph.D theses, books) present in the literature have been analysed. In fact, most of the relevant works have been produced by scientists at three universities: the research branch in marine modelling and simulation at Norwegian University of Science and Technology (NTNU), Delft University of Technology (TU Delft), and Genova University (UNIGE).

In particular, Fossen from NTNU wrote three comprehensive books [23–25] about the equations of motion and propulsion control systems for ocean vehicles. In these books, the mathematical basis and the main steps to study ship motion are well explained. Fossen also dedicated some detailed sections for control systems of ship motion. Two Ph.D theses developed in NTUN have been considered [33, 40] that deal with the modelling of propeller hydrodynamic forces and the local control system. In particular, the latter is focused on the propeller thrust estimation in four-quadrants operations.

In the literature are also several works about ship manoeuvrability. In particular, Ankudinov [11] proposed a modular mathematical approach for real time manoeuvring simulation; Kijima [30], studied roll motion; Brix [18] wrote regarding manoeuvring technical problems.

With regard to the rudder tests data, an active research group is located in Southampton; several works have been produced [31, 32] in which the rudder behaviour downstream of the propeller has been studied, proposing different mathematical formulas to evaluate interferences between the hull and the propeller.

The most studied topics include ship propulsion, the main engines, and auxiliary systems. However, most of the works currently in the literature deal only with particular propulsive components or with a particular ship system without taking into account the behaviour of the whole ship. Propeller pitch change mechanisms have been thoroughly studied in several works [12, 29, 44] from the point of view of mechanics, hydraulic system, and acting loads.

Gas Turbine dynamics analysis can be found in some works [14, 35, 37], and of particular note, the works of Rana and Rubis also deal with the turbine control system.

Few studies on the ship simulation, including propulsion plant, manoeuvring model, and control system, are reported in literature [16, 38].

In the present thesis, the propulsion simulation of a particular CODLAG system will be addressed and for which no information is present in literature as only three of these systems have been installed (on the Finnjet ferry, on the German Navy F125 frigate, and on Italian-French Navy Frigate FREMM). The modelling and simulation of such systems have been popular topics at Genoa University for the past fifteen years. Different propulsion plant simulation models (ferries, merchant ships, pleasure boats, and navy vessels) have been developed and published [5, 7, 13, 15]. In particular, the Cavour project [2, 3, 6] has been the best example of the effectiveness of simulation techniques.

2.3 Ship Dynamics

The simulation model will be a software platform that allows the study of the vessel behaviour during transient conditions (acceleration, deceleration, etc) and during steady state conditions (constant speed navigation) as well as the analysis of the mutual interaction between all the elements involved. To reach this goal, three completely different ship macro systems were joined together to adequately describe the global ship behaviour: the ship manoeuvrability, the propulsion plant, and the control system.

Each macro system is composed of different elements; each of these elements have been schematized and modeled using the differential equations that govern their physical behaviour and represent their functions.

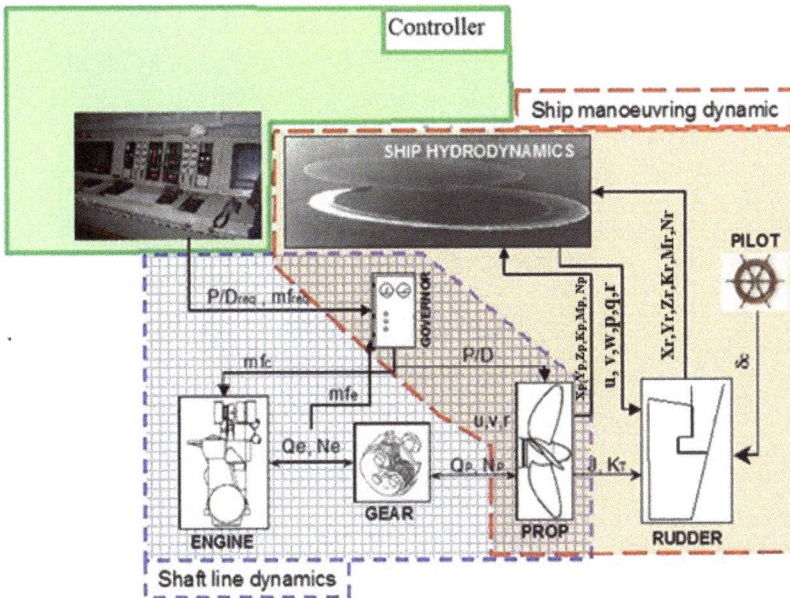

Figure 2.1: Ship Simulation Platform

The subdivision adopted in Table 2.1 is not completely accurate, because it is impossible to define the element membership to a specific macro area, and some systems are borderline. For example, the propeller can be seen associated with the manoeuvrability as it provides the thrust required to achieve a given ship speed, but it can also be included in the plant propulsive system because it determines the engine load.

The different sub models have been studied and developed with different degrees of detail. The simplest way to schematize a system is by using a table of parameters and algebraic equations that identify the system behaviour in steady state conditions.

Table 2.1: Simulator scheme

MACRO - SYSTEMS	SUB - SYSTEMS
Manoeuvrability	Hull
	Rudder
	Appendages
	Propeller
Propulsion Plant	Main Engine
	Gearbox
	Shaft line
Control System	Propulsion Control System

Usually most of the needed data comes from system manufacture trials. The complex and more realistic approach, used for the most of the elements, is to model the system with its physical equations, both algebraic and differential. With this approach, a large set of parameters must be taken into account, and this itself can be a difficult obstacle.

An intermediate approach could be obtained by merging tabular models and algebraic/differential equations to produce a quasi-static model.

Obviously, it is important to take into account the computational cost that increases (more than linearly) with the model complexity (Fig. 2.2).

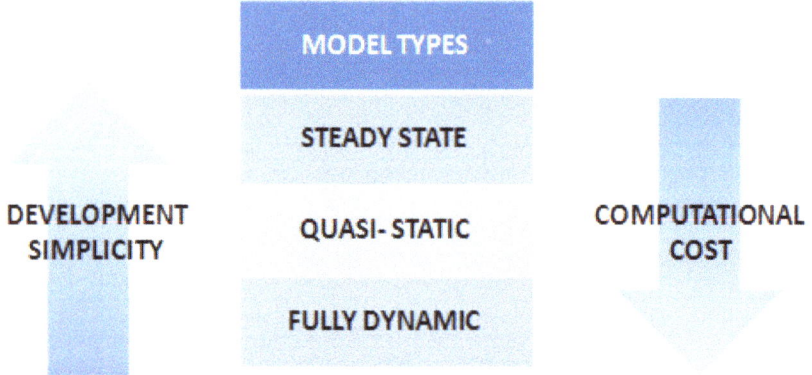

Figure 2.2: Simplicity vs. computational cost

2.3.1 Rigid Body Dynamics

The first macro system mentioned was the manoeuvrability, as the aim is to study the vessel's motion as a rigid body with six degrees of freedom (surge, sway, heave, roll, pitch, and yaw). In order to study the vessel dynamics, some simplifications need to be introduced: the ship is considered to be a rigid body; a constant displacement is assumed, so it is possible to apply the Newton-Euler formulation as follows:

$$(M_{RB} + M_A)\,\dot{v} = C_{RB}(v)v + D_L v + D_{NL}(v)v + G(\eta)\eta + \tau_P + \tau_R$$

Where:

M_{RB} and M_A are the inertia and added mass matrices, respectively

C_{RB} and G the Coriolis and restoring mass matrices, respectively

D_L and D_{NL} are the linear damping and non linear damping matrices, respectively

τ_P and τ_R are the propeller and rudder forces and moments vectors, respectively

$\dot{v}$ is the acceleration vector

v is the velocity vector

η is the position vector

All the elements that appear in the motion equation are explained in detail in Chapter 4.

The crucial task has been to evaluate the right side of the equation. In particular, the hydrodynamic forces are the most difficult to evaluate with a standardized methodology valid for each vessel type.

The rigid body differential equation has been developed in matrix form and gives information about the ship acceleration, velocities (both linear and rotational), and ship position in terms of trajectory and attitude angle.

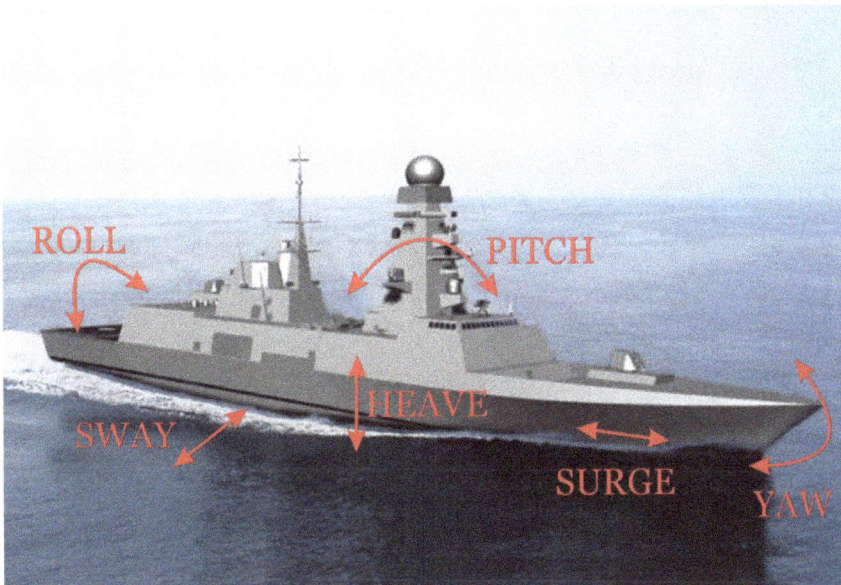

Figure 2.3: Six degrees of freedom

2.3.2 Propulsion plant dynamics

The propulsion plant is the second macro system studied, and its contribution to the global ship behaviour is considerable.

It produces the required thrust, with the highest efficiency possible, to overtake hydrodynamic drag. The propulsion plant has been decomposed into the following main elements: main engines and their governors, the gearbox, the thrust and other bearings, the shaft line, and the two controllable pitch propellers.

A large number of variables are required to represent the propulsion plant dynamics, including ship speed, rpm, propeller pitch angle, and fuel supply. In a modern propulsion plant, all these variables are managed by the propulsion control system.

It is hypothesized that the propulsion plant dynamics are mainly affected by the engines dynamic and the transmission lines dynamic; thus, both have been taken into account.

The main engines (i.e. gas turbine, electric motor, and diesel engine) dynamics can be charged to their thermodynamic characteristic and to the electronic control effects.

The main engine modelling and the main assumptions employed are reported in Chapter 3.

The transmission line studied in this work has two degree of freedom: the shaft line revolution regime and the propeller pitch angle. With respect to traditional transmission lines, this gives a greater operational flexibility.

The dynamics of the propeller pitch is governed by the load acting on the propeller blade and by the dynamics of the pitch actuating mechanism. For the last mechanism, a detailed mathematical model has been developed to include both the blade hydraulic piston motion equation and the dynamic pressure change inside the piston actuating chamber. To evaluate the loads acting on the propeller blade, a rigorous formulation has been developed.

The two motion equations are reported herein.

$$\ddot{\varphi}(t) = \frac{1}{I_b}\left(Q_{hyd}(t) + Q_S(t) + Q_{-\varPhi}(t)\right)$$

Where:

$\ddot{\varphi}$ is the blade angular acceleration

$Q_{-\varPhi}$ is the torque due to the interaction forces between propeller blade and blade bearing

Q_{hyd} is the hydraulic torque

Q_s is the total spindle torque acting on the blade

I_b is the moment of inertia of the blade about the spindle axis f_{-3}

$$m_m \ddot{x}_{pist}(t) = A_1 p_1(t) - A_2 p_2(t) - B_p \dot{x}_{pist}(t) + \sum_{i=1}^{z} \varPhi_i(t)$$

Where:

m_m is the total mechanism mass

$\ddot{x}_{pist}$ is the piston acceleration

A_1 and A_2 are the yoke areas of the astern chamber and of the ahead chamber, respectively

p_1 and p_2 are the pressures inside the two chambers

B_p is the damping coefficient

$\sum_{i=1}^{Z} \Phi_i$ are the interaction forces between each blade and the bearing

Z is the blade number

To evaluate the inside pressure, the mechanism is proposed in the following formula:

$$\dot{p}_{oil}(t) = \left(q_i(t) - C_{ip}p_{oil}(t) - A\dot{x}_{pist}(t)\right) \frac{B}{V(t)}$$

Where:

p_{oil} is the pressure

q_i is the volumetric flow in

C_{ip} represents the leakage coefficient

Agg is the piston area

Bgg is the oil Bulk modulus

$V(t)$ is the chamber volume

A detailed explanation of the propeller load and the propeller pitch actuating mechanism behaviour is reported in Chapter 6.

The second degree of freedom of the system is the shaft line rotational regime.

The shaft line has been modelled using the Lagrangian equation, which provides the rotational propeller speed ω.

$$I_s\dot{\omega}_s(t) = Q_{eng}(t) - Q_{fric}(t) - Q_P(t)$$

Where:

I_s is the transmission line polar inertia

$\dot{\omega}_s$ is the shaft line acceleration

Q_{eng} is the engine torque

Q_p is the required propeller torque

Q_{fric} is torque due to friction

The shaft lines equation is a scalar equation, and, once solved, gives information about the acceleration and revolution regimes of the shaft lines. It is also possible to know the instantaneous position of the propeller blade. For a comprehensive description of the shaft line model and behaviour, see Paragraph 3.2.

In the case of a twin-propelled ship, the two shaft lines are considered independently; thus, two equations should be implemented, one for each shaft line.

This becomes necessary because, during ship manoeuvring, an asymmetrical propeller behaviour is experienced. In this implementation, the system is being represented in a more realistic way.

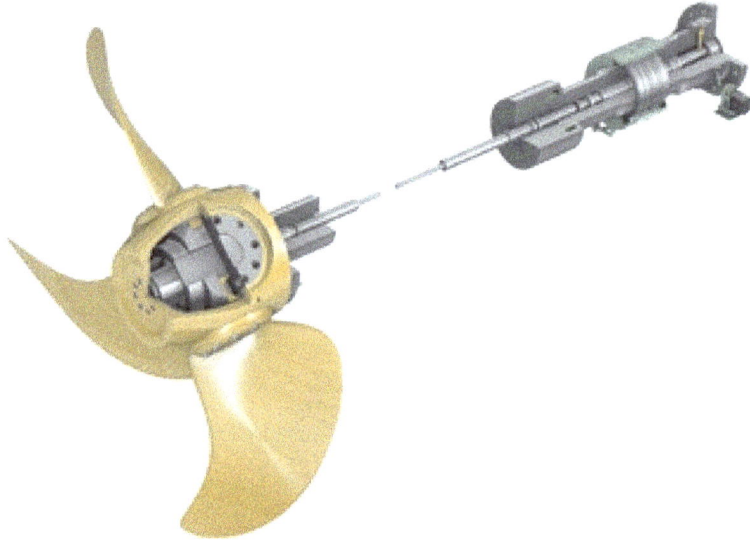

Figure 2.4: Shaft line overview

2.3.3 Control system

All modern ships have installed electronic controllers in order to assist the crew in handling the ship safely and efficiently.

A modern control system is a fully integrated system (ship supervisor) covering many aspects of the ship operation that include the propulsion plant operation, power management operation on the auxiliary engines, auxiliary machinery operation, cargo on-and-off-loading operation, navigation, and administration of maintenance and purchasing of spare parts.

Within the large group that composes the ship supervisor, in this thesis, only the propulsion control system has been studied and designed.

The aim of the propulsion control system is to '*translate*' the operator will into a suitable machinery signal (setpoint) respecting all constraints (torque max, rpm max, fuel consumption and so on) for all navigation modes.

Most of the parameters are controlled by a P.I.D. algorithm; the command signal (output) is obtained by comparing operator requests with the data from the various models, giving the propulsion plant a well-defined dynamic.

An example of P.I.D. algorithm is shown below:

$$u\left(t\right) = K_P e(t) + K_I \int_0^t e\left(t\right) dt + K_D \frac{d}{dt} e(t)$$

Where:

$u(t)$ is the setpoint
$e(t)$ is the error between the reference and the actual value
K_P is proportional gain
K_I is integral gain
K_D is derivative gain

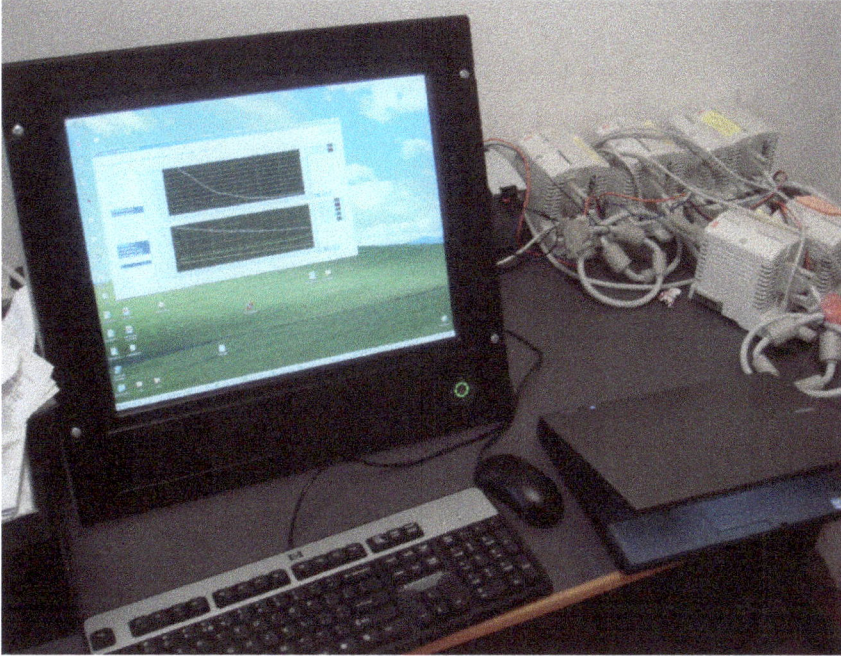

Figure 2.5: Example of propulsion controllers

A combination of P.I.D. algorithms are used, for instance, to generate the setpoint for the two propulsion degrees of freedom (shaft line revolution regime and propeller pitch angle).

The control system dynamics is not only supported by the P.I.D. algorithm; it is also influenced by a set of logics, thresholds, Boolean state, and truth table. For example, two innovative logics for emergency manouevring have been developed: *Slam Start* and *Crash Stop*.

The combination of all these elements produces the propulsion control system dynamics.

For the complete description of the ship propulsion control system, see Chapter 4.

2.4 System of Systems

The peculiarity of this work, as previously announced, is a system engineering approach to bring together all disciplines involved to represent a unified view of the system. This creates the need to implement many differential equations. From a mathematical point of view, the problem can be summarised by solving the following second order differential equations system (only the main equations are reported):

$$
\begin{cases}
(M_{RB} + M_A)\,\dot{v} = C_{RB}(v)v + D_L v + D_{NL}(v)v + G(\eta)\eta + \tau_P + \tau_R \\
\ddot{\varphi}(t) = \dfrac{1}{I_b}\left(Q_{hyd}(t) + Q_S(t) + Q_{-\phi}(t)\right) \\
m_m \ddot{x}_{pist}(t) = A_1 p_1(t) - A_2 p_2(t) - B_p \dot{x}_{pist}(t) + \displaystyle\sum_{i=1}^{Z} \Phi_i(t) \\
\dot{p}_{oil}(t) = \left(q_i(t) - C_{ip} p_{oil}(t) - A\dot{x}_{pist}(t)\right)\dfrac{B}{V(t)} \\
I_s \cdot \dot{w}_s(t) = Q_{eng}(t) - Q_{fric}(t) - Q_P(t) \\
u_i(t) = K_P \cdot e(t) + K_I \cdot \int e(t)\,dt + K_D \cdot \dfrac{d}{dt} e(t)
\end{cases}
$$

Solving this system is difficult and time consuming due to the large number of variables involved. This system was developed into a software environment able to solve it in the time domain. Thanks to modern computational power, this system can be solved an infinite number of times in order to predict the ship dynamics with different initial and boundary conditions.

Another no less important 'system' included to the simulation platform is the human factor; this kind of 'system' cannot be mathematically expressed. Incorporating the human interaction into the simulation loop is essential to have more realistic predictions about the ship dynamic behaviour under real conditions.

In Fig. 2.6 the connection between all simulation platform macro systems and elements are shown. On top, the human figure that handles the system is addressed. The control system 'translates' and processes the input into a working point for the machinery. The propulsion plant allows the system to achieve the human request.

In order to better understand the figure, each sub-system is represented by black circles and the straight lines represent mutual interactions.

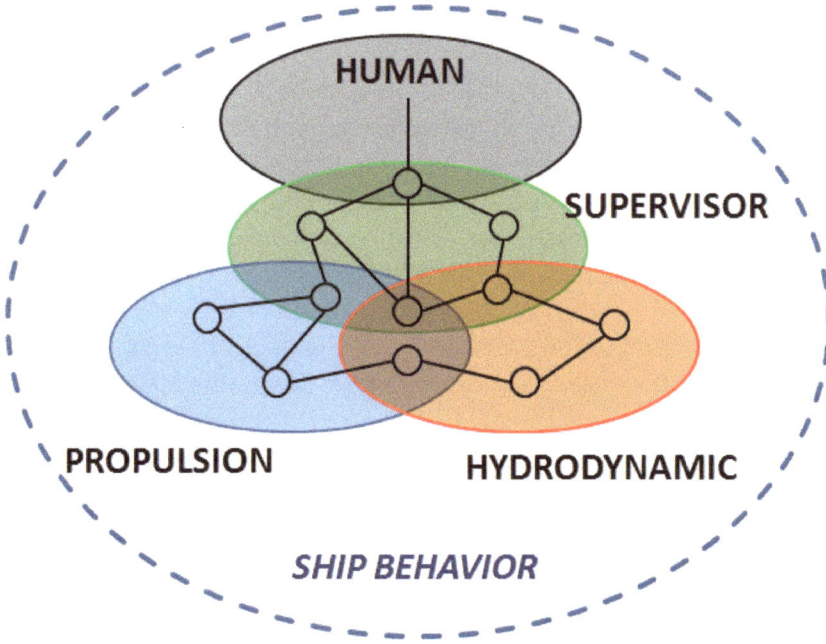

Figure 2.6: System of systems

3 Ship Propulsion Plant

The propulsion plant, the objective of this study, is a CODLAG (COmbined Diesel eLectric And Gas) system. In a CODLAG system, as seen in Fig. 3.1 Propulsion plant, electric motors are used at low speeds, the gas turbine is used at medium-high speed, and both prime movers are used to reach maximum speed or maximum power, if needed. With this kind of combination, a ship has more flexible power availability.

The most power is guaranteed by a gas turbine, which is linked to the input gearbox shaft. The electric motors, fed by diesel generators, are directly mounted on the shaft lines. A particular kind of gearbox (X-cross) is present: one input, double output, with a double reduction stage. Two shaft lines drive two controllable pitch propellers.

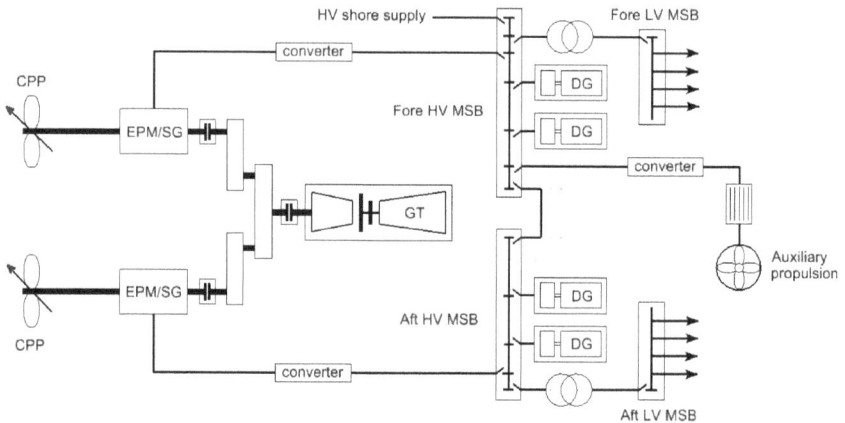

Figure 3.1: Propulsion plant

A Summary of the main elements present:
- One gas turbine
- Two electric motors
- Four Diesel generators
- One gearbox
- Two controllable pitch propellers

3.1 Main Engines

3.1.1 Gas Turbine

The choice of a suitable gas turbine model is strongly dependent on the particular system under investigation. In the first approach, the gas turbine mathematical model has been structured in a modular arrangement, in which each module, representing a specific engine component (i.e. compressor, high pressure turbine, power turbine, combustor, etc.), is modelled by means of steady state performance maps, time dependent momentum, energy, and mass equations, and non-linear algebraic equations [14].

The main differential equations to be solved in the time domain concern both the mechanical and thermodynamic engine parameters. In particular, the working fluid processes are governed by the continuity equation:

$$\frac{d(\rho V)}{dt} = M_i - M_o$$

and by the energy equation:

$$\frac{du}{dt} = \frac{1}{\rho V}(M_i h_i - M_O h_O) + M_f LHV$$

Where:

M_i is the inlet mass flow rate
M_O is the outlet mass flow rate
V is the GT component volume
h is the enthalpy
u is the internal energy
LHV is the fuel lower heating value

The dynamics of the rotating shafts, both for high and low pressure turbines, are determined by the dynamic momentum equation:

$$\frac{d\omega}{dt} = \frac{1}{J}(Q_m - Q_{br})$$

Where:

Q_m and Q_{br} are the motor and brake torques
ω the angular velocity

Unfortunately, the introduced mathematical equations for the numerical modelling of marine gas turbines are not usually very reliable at very low engine loads.

In order to overcome this difficulty, more effective methods have been developed based on engine manufacturer trials using the turbine maps.

The engine power can be simulated by a numerical surface depending on engine speed and fuel consumption flow rate. In particular, in Fig. 3.2, the power surface of a gas turbine is reported. The surface has been modelled on the ground of the load diagram provided by the engine manufacturer. Using these data, it is possible to obtain the gas turbine power (or torque) through a surface interpolation. The fuel flow is evaluated by the gas turbine controller that transforms the rpm set-point into a fuel flow set-point.

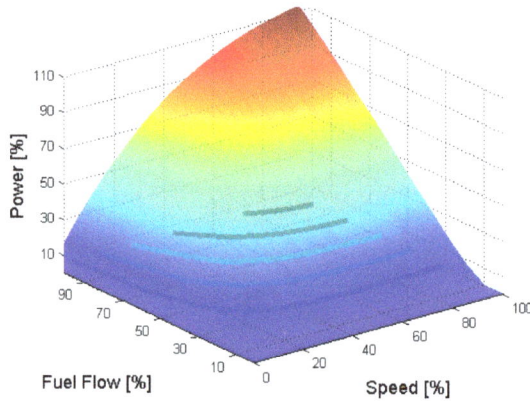

Figure 3.2: Power surface of the gas turbine

In the gas turbine model, a sub-model that represents the inner machinery governor has also been included known as the Turbine Control System (TCS). The TCS is in the second hierarchic level loop. The main controlled variables are, in analogy with the propulsion control supervisor: the torque, the regime of revolution, temperature, and the fuel flow [1].

3.1.2 Electric Propulsion Motor

Electric motor models used for the propulsion system analyses can be grouped into two types of models: non-physical models and physical models.

By the first approach option, which is employed in this work, the engine is modelled by simple equations to set up the electric propulsion motor behaviour; thus, it is necessary to identify few parameters from the engine manufacturer data. By this technique, only the torque and the rpm of the engine are represented without any reference to the machine electric variables. This approach is suitable only for a high-level simulation.

The torque is calculated according to a proportional and integral action on the speed error [8]. In the simulation process, the engine actual speed is compared with the engine speed commanded by the governor, and then the engine torque is adjusted by a non-linear action in order to achieve and maintain the desired engine speed.

The torque and over-speed protection are represented in order to ensure a more realistic behaviour.

3.2 Transmission Line

Due to the axial symmetry of the transmission line (gearbox, shaft, and propeller) and neglecting the shaft line deformation, the system can be studied as a holonomic system with one degree of freedom: the shaft line angle.

The shaft line, shown in Fig. 3.3 Shaft line dynamics, has been modelled using the Lagrangian equation:

$$\frac{d}{dt}\left(\frac{T}{\partial\dot{\Phi}}\right) - \frac{T}{\partial\Phi} = Q$$

Where:

Q is the stress acting on the system

$T(t)$ is the kinetic energy

Φ is the system degree of freedom

The stress Q is the sum of the moments acting on the transmission line.

$$Q(t) = \sum_{i=1}^{n\,eng} Q_{eng}(t) - Q_{fric}(t) - Q_P(t)$$

Where:

Q_{eng} is the engine torque

Q_p is the required propeller torque

Q_{fric} is the torque due to the friction

n_{eng} represents the engine number

The required propeller torque is evaluated using the procedure illustrated in Paragraph 6.4.2; the torque due to friction is evaluated through a table representing experimental data as a function of regime of revolution and delivered power; the engine torque evaluation is shown in Paragraph 3.1.

Since the system is considered as a rigid body with a fixed axis, the total kinetic energy is the algebraic sum of the kinetic energy of each component:

$$T = \frac{1}{2}\omega I_G(\omega) = \frac{1}{2}\left(\sum_{i=1}^{n} I_i\omega_i^2\right)$$

Substituting the expression of Q and T into the Lagrange equation, expressed in terms of shaft line regime of revolution, we obtain:

$$I_s\dot{\omega}_s(t) = \sum_{i=1}^{n\,eng} Q_{eng}(t) - Q_{fric}(t) - Q_P(t)$$

Where:

I_s is the transmission line polar inertia

$\dot{\omega}_s$ is the shaft line acceleration

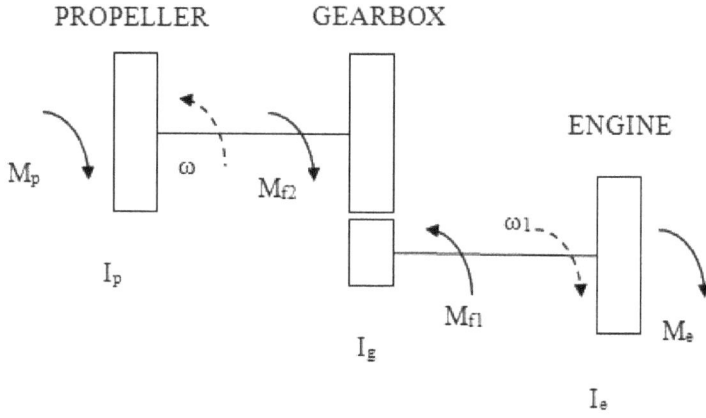

PROPELLER   GEARBOX

ENGINE

M_p ω M_{f2}

$\omega 1$ M_e

I_p

M_{f1}

I_g

I_e

Figure 3.3: Shaft line dynamics

4 Propulsion Control

The increasing demand for fast and economical marine transport of passengers and freights is giving renewed interest to several innovations and developments which will set the new standards for the next ship designs.

Major innovations have already focused on the propulsion systems, and they have been designed to satisfy the mandatory rules for safety and environmental impacts and to reduce fuel oil consumption. Moreover, the increasing complexity of these new marine propulsion systems, which often include engines with different power levels and/or of different engine types (especially in navies) leads necessarily to the development of dedicated propulsion control systems that are able to manage such complex system in safe and economic ways for various vessel conditions. In some cases, the propulsion control systems must also be able to ensure top performance of the propulsion plant.

Two different techniques are usually available to develop and design a control system: the open loop and the closed loop.

The open loop control system is composed of the controller (PLCs) and the transmission lines and machinery. By this approach, the variable to be managed is set to a specific value, but since there is no comparison with the field value, it is not certain that the controlled variable assumes the desired value.

In the closed loop control system, the main elements of the system are the controller (PLCs) and the transmission lines, machinery, and sensors. The controlled variable is regulated (set-point) by a continuous comparison with the data coming from the field (feedback). Usually, in marine propulsion control system design, the closed loop is preferred because it is more accurate. The main reason is the fact that the ship operates in a hostile environment, the sea, in which the environmental conditions may vary quickly with a magnitude that cannot be predicted. Using the closed loop approach, information about performance is measured and used to correct the system behaviour.

In order to test performance and reliability of the control logics, new design approaches for automation development have been recently introduced [4]; one such example is the "Real Time Hardware in the Loop" (RT-HIL).

According to this design approach, the propulsion control system is first designed and tested in a virtual environment, and then the control logics are implemented into a PLC linked to a virtual ship model to evaluate the effectiveness before onboard installation. This approach provides the designer realistic feedback [2].

4.1 Propulsion Control System

The supervisor system of the ship is the link between human action and the machinery onboard. The crew can interact with the ship through operation stations, linked to each other by a LAN that forms the control network. At the end of the control network, automation servers trough the I/O cards between the control and the field bus networks. These elements are the joints that transform the human action (the press of a button, the pull of a lever) into a proper signal to the PLCs. In addition, incoming field information is stored in these servers.

The field bus network is composed of the PLCs and the sensor needed to manage the different systems. In this network, all data coming from the machinery are monitored such that the PLCs can then provide appropriate set-points to each element of machinery.

For safety reasons, the LAN infrastructure is usually a double ring for both control and field bus networks..

The ship supervisor is usually divided into different sub-systems as is shown in Fig. 4.1. Each sub-system manages and controls different onboard activities: ship safety, propulsion, electric power management, etc. The focus of this work is the design of the particular characteristics of the propulsion control system.

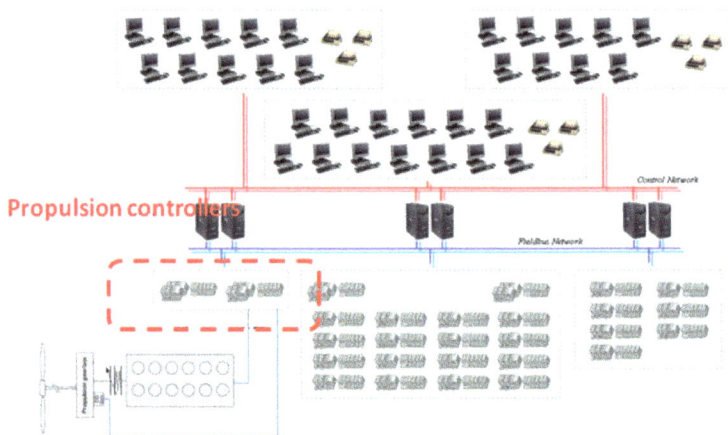

Figure 4.1: Ship Supervisor Overview

The propulsion control system can be designed with different hierarchic levels as shown in Fig. 4.2.

The higher level (indicated in black lines) is the ship supervisor, developed by the shipyard or by an automation provider; its tasks are to transform human action into

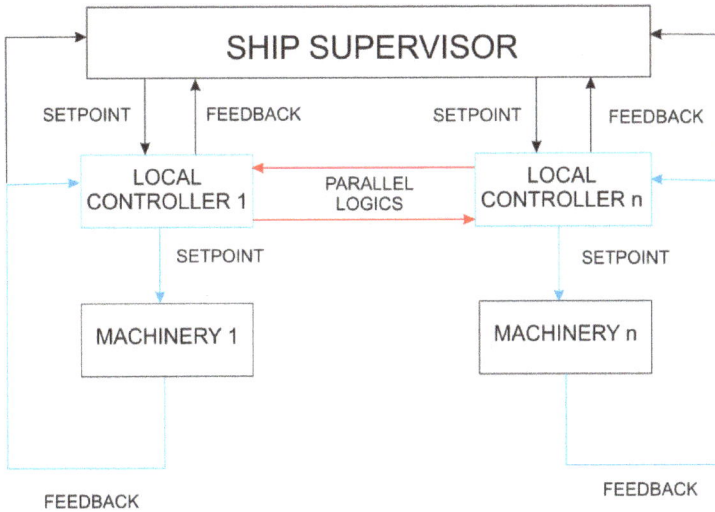

Figure 4.2: Hierarchical Levels

a set-point signal to the machineries and prevent possible system failure (i.e. due to exceeding the maximum speed or torque capacities, high pressure, etc.).

Next, the second level (indicated in blue lines) is composed of the local controllers given by the machinery manufacturers; the local controller receives as input the set-point coming from the supervisor and returns a signal that confirms the receipt. The local controller includes the control logics needed to preserve the machinery safety, including thresholds of the imposed limits that are higher than those allowed by the supervisor controller. Finally, the local PLC sends the proper signal, eventually modulated by the inner logics, to the machinery.

A third level also exists called 'parallel control logics' (indicated in red lines); in this level, the single local controller can interact with the other local controllers and machineries without the supervisor control. Usually, due to confidential reasons, these logics are not well known to the supervisor designers and some unexpected behaviours can occur. These behaviours are obviously dangerous because:

"The controller mustn't lose control".

The challenge is to design the control logics and to set the main ship supervisor parameters to reach the desired performance in a safe and economic way. Another aim is to set all the propulsion control system thresholds to avoid, as much as possible, the intervention of the local controllers and parallel logics. The latter task can be difficult without a full collaboration between the automation provider and the machinery manufacturers.

In more detail, the crew interact with the propulsion plant from the bridge or from other command stations onboard. The standard practice is to manage the ship through a control approach based on propeller speed management. For complex propulsion systems, it is also possible to change the propulsion mode (i.e. navigation, harbor, etc.). The desired speed can be achieved through operation in two different ways:
- Through step levers in the 'automatic' way.
- Setting the rpm and/or the propeller pitch set-points independently and in manual mode.

We would like to analyse and model the first case. In designing a propulsion control system we should keep in mind two different purposes: regulation and protection. Regulation allows the system to reach desired performance; protection allows the system to do this safely.

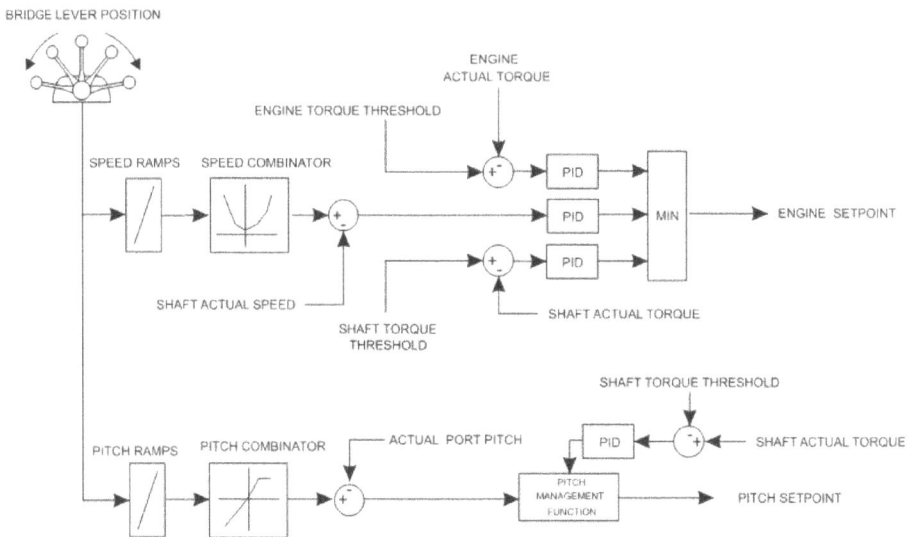

Figure 4.3: Propulsion Controller layout

In Fig. 4.3, the logics flow to manage the two propulsion plant degrees of freedom (the propeller pitch and the shaft line revolution) is shown. The input signal coming from the bridge lever position is actuated by a human action. The signal is split into two signals, one for each degree of freedom. The signals are then modulated by auto-adapting ramps. The modulated signal is converted to a set-point (speed or pitch), and here ends the 'regulation' control loop. This first part is very similar to a standard open loop. Subsequently, the signal is compared with the field values and therein begins the 'protection' part of the control system that ensures the safety of the system. Here, all protections are implemented (explained in detail below). It is thus possible to consider the propulsion control system as a hybrid between an open and closed loop. When shipbuilders or ship-owners speak of ship control in terms of speed, the indicated numbers are by definition not exact because, as external conditions (sea, wind, current, performance degradation of propeller and hull) change, the desired speed may not be reached!

4.1.1 Ramps

The ramps are a set of constants defined for each value reached by the step lever. The ramps modulate the signal coming from the lever of the telegraph, which is usually a step. The ramps are different for pitch and rpm, for acceleration and deceleration, and also for the various propulsive modes. In Fig. 4.4, the set of rpm ramps in navigation mode for an accelerating manoeuvre is shown in non-dimensional form.

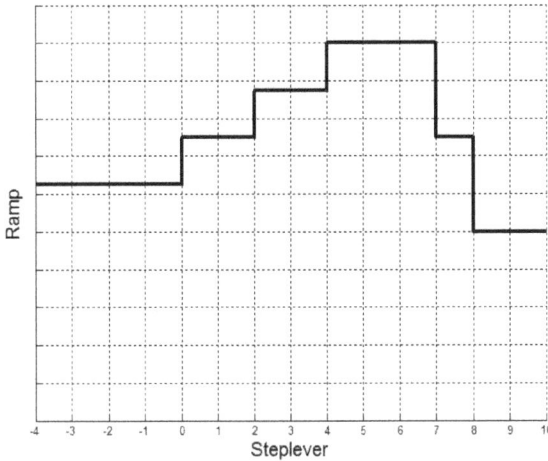

Figure 4.4: Example of ramps

The value for the ramps is the angular coefficient, which is assigned to the time variant signal incoming to the block (the unit of the constant ramps values are 1/s or Hz). In this way, the signal that arrives to machinery is more gradual and thus avoids the possibility of receiving signals with large discontinuities, which may create problems for the PLCs. In Fig. 4.5, the ramps block effects in the time domain are shown.

The study of these values has been possible thanks to an intense simulation campaign through which the appropriate values have been selected and tested. Values are considered appropriate such that during tight manoeuvres, the machineries are able to avoid heave cuts due to control system protections.

All these values are implemented into a proper block within the controller structure.

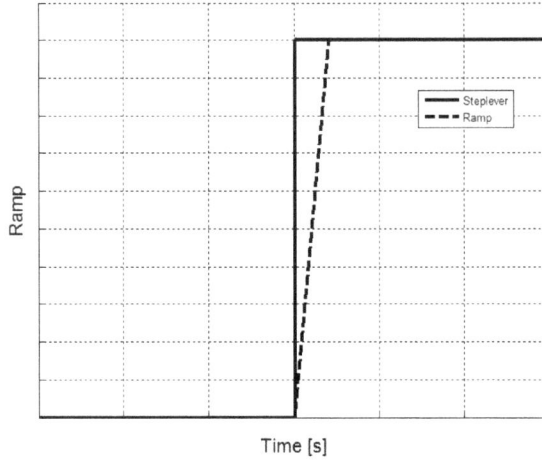

Figure 4.5: Effect of ramps vs. time

4.1.2 Combinator Law

The ship speed regulation is performed by modulating the propeller rpm and/or pitch. The performance variation rules and hence the possible combinations of rpm/pitch values (combinatory curves, depending on lever position) have been studied in steady state conditions and then translated into a software program implemented in the main propulsion controller according to the set mode of the vessel. The combinatory curves, an example of which is shown in Fig. 4.6, provides the optimum performance of the ship in the various propulsion modes. With regard to this important topic, it is appropriate to note that the combinatory curves are studied to minimize fuel consumption [22].

The signal, modulated by the ramps, is converted into a reference value for the propeller pitch and shaft line revolution regimes by means of two tables in which the steady state equilibrium points for propeller pitch and shaft line rotation regimes (depending on lever position) are reported.

During normal management, when all controlled variables are within the allowed limits, the set-point both for rpm and pitch is generated by a P.I.D algorithm based on the error between the desired value and the monitored one:

$$e\,(t) = theresold\,(t) - feedback(t)$$

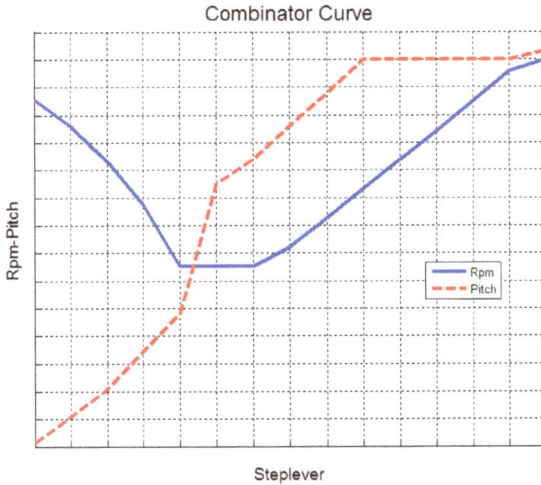

Figure 4.6: Example of combinatory curves

$$setpoint\,(t) = K_P e(t) + K_I \int e\,(t)\,dt + K_D \frac{d}{dt}e(t)$$

4.1.3 Protections

Different protection logics can be applied to a propulsion control system as a function of the propulsion plant. Herein the main logics studied are reported:
- Engine over-torque protection
- Engine over-speed protection
- Shaft line over-torque protection
  - Rpm reduction
  - Pitch reduction
- Shaft line over-speed protection
- Torque balance

Obviously, other variables are monitored by the ship supervisor and become part of the protection loop, including the main engine temperatures (inlet air, exhaust gas, coolant), gearbox oil pressure and temperature, thrust bearing oil pressure and temperature, propeller thrust, etc.).

The required propeller speed is compared with the measured actual shaft speed to feed a PID algorithm able to assess the main engine throttle demand. This kind of speed error is adjusted to keep each engine within its torque limit during every condition to prevent overload and also to take into account the other variables influenced by the shaft line rpm. The final engine throttle is the minimum signal between the results achieved by three PID algorithms, which act:

- On the shaft speed error;
- On the difference between the engine torque limit (for instance, equal to 85% of the main engine nominal torque) and the engine actual torque;
- On the difference between the shaft torque limit (for instance, equal to 85% of the design shaft torque) and the shaft actual torque.

Once the engine throttle demand is calculated, the corresponding local engine control system must regulate the fuel flow to achieve the proper power required by the propeller.

The pitch set-point is possibly corrected to keep each shaft line within its torque limit during every condition. A possible correction to the pitch set-point is made through a P.I.D. algorithm on the basis of a torque limit which is lower than the previous one (i.e. 75% – 85% of the shaft nominal torque). Possible overload protection is determined first by the pitch reduction and then by the main engine throttle reduction.

During manouevres for a twin-propelled vessel, the loads on the propellers are different between the inner and the outer shaft line, so different torques are acting. In the case of x-crossed gearboxes, this difference in acting torques can be dangerous for gearbox teeth, so a proper control function, referred to astorque balance, was designed. The pitch set-point is corrected for each shaft on the basis of the difference of torques acting on the shafts. The pitch set-point is then reduced for the shaft subjected to the higher torque.

It is difficult to formulate a mathematical expression for the complex control propulsion system under investigation, but it is possible to express such an arrangement as a functionality dependence:

$$n_{set}(t) = f(V, Tel, Q_{stbd}, Q_{port}, Q_{eng}, nav\ mode, n_{max}, n_{min}, n)$$

$$\varphi_{set}(t) = f(V, Tel, Q_{stbd}, Q_{port}, n, nav\ mode, \varphi_{max}, \varphi_{min})$$

Where:

V is the ship speed

Tel is the step lever position

Q_{stbd} and Q_{port} are the real starboard and port shaft torques, respectively

Q_{eng} is the real engine torque

$nav\ mode$ is the real operative mode

$n_{max} and n_{min}$ are the maximum and the minimum shaft line revolution regimes, respectively

$\varphi_{max}\varphi_{min}$ are the maximum and the minimum propeller pitch angles allowed

n is the actual shaft line regime of revolution

4.2 Emergency Manoeuvres

Undoubtedly, propulsion systems with controllable pitch propellers serve an advantage for emergency manoeuvres. The possibility of varying two degrees of freedom (propeller pitch and rpm) ensures a much faster ship response. For example, in a crash-stop manouevre, the time and the space needed to arrest the ship hugely decreases if compared to a propulsion plant with a fixed propeller. It is not required to stop and reverse the shaft line motion; rather, the propeller pitch only needs to be changed. During rapid acceleration, it is possible to increase both the shaft line speed and the propeller pitch. These great benefits are offset by complex control logic studies investigating a system with two degrees of freedom.

In the following discussion, the control logics, developed by the author, are designed to maximise the performance in the two cases outlined above.

4.2.1 Slam Start

For the proposed control logic, the automation system interprets as a "Slam Start" condition every ship manoeuvre in which the bridge lever position is suddenly brought to the maximum step (100%), starting from a step value lower than 70%. In this context, "Slam Start" manoeuvres can also start when the vessel is moving astern. In this case, propeller pitch and speed are managed according to a proper combinatory law to drive the CPP until the zero thrust condition. On the contrary, if the initial pitch is greater than that one corresponding to zero thrust, φ_0, shaft speed is managed by a PID action on the basis of the speed error between actual speed and commanded set-point. The commanded set-point provides the maximum propeller speed modulated by a ramp over time.

Using simulation techniques, great numbers of 'virtual manoeuvres' can be calculated to optimize the slope of the ramps in a very short time with respect to traditional methods.

The propeller pitch set-point is calculated over time by a proper control function as described in Fig. 4.7.

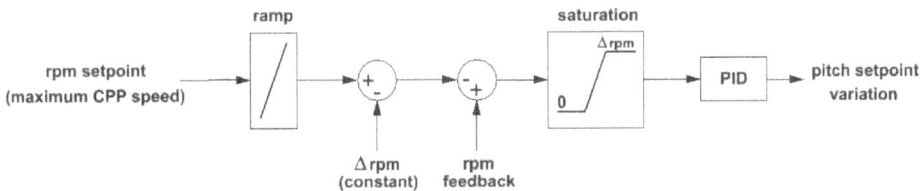

Figure 4.7: Propeller pitch set-point logics in the "Slam Start" manoeuvre

The whole control logic is summarised by means of the flow chart illustrated in Fig. 4.8.

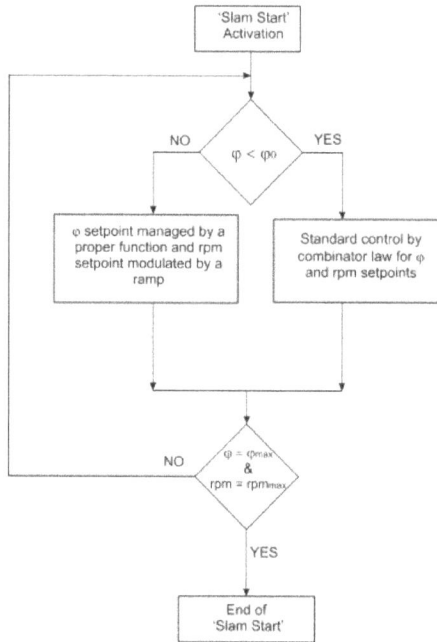

Figure 4.8: Flow chart for Slam Start control logic

In particular, when the propeller pitch increases the torque required by the propeller very quickly, occasionally the main engine is not able to deliver the necessary torque; this causes a drop in rpm and a slower manoeuvre. To prevent this, the proposed pitch control function acts on the propeller pitch taking care that the actual propeller rpm follows its increasing set-point with a speed tolerance (Δrpm) in a smooth and linear way and for as far as possible. In addition, in this case the simulation technique permits the set, as well as possible, the speed tolerance ((Δrpm) value.

The speed tolerance (Δrpm is useful to activate the "*pitch freezing*" when the shaft line rpm is very far from its set-point; otherwise, any increase in the pitch set-point would result in a further drop of shaft revolutions.

This particular behaviour is shown in Fig. 4.9, where the propeller pitch set-point (solid line) and the corresponding feedback (dash-dot line) are reported.

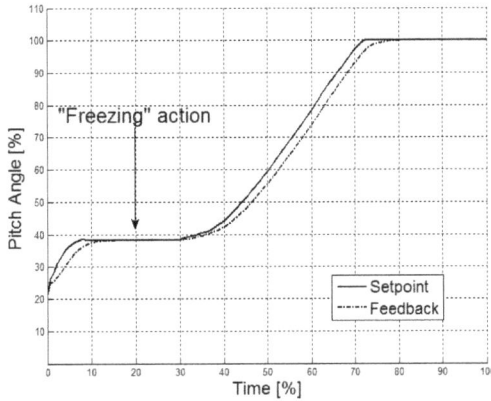

Figure 4.9: Propeller pitch dynamics for "Slam Start" manoeuvre

4.2.2 Crash Stop

For the "Crash Stop" condition, similar to that of the "Slam Start" condition, several possible critical manoeuvres are considered, and this aspect complicates significantly the control optimization for the automation designer. In fact, every condition in which the bridge lever position is moved from any step higher than 20% to an astern step at least equal to –100% is considered a "Crash Stop" mode. In order to reduce the parameter range to be set for any possible case, the following logic has been studied (Fig. 4.10).

As long as the propeller pitch values φ are greater than pitch corresponding to zero thrust (φ_0), the pitch set-point is calculated by a proper function (based on a PID action on the error between actual propeller rpm and a rpm threshold set to a constant value), whereas the shaft speed is managed acting on the engine signal responsible for the engine fuel flow calculation. During a crash stop manoeuvre, the quick decrease of the pitch propeller causes an increase in shaft line revolution with the risk of a non-controllable over-speed due to the windmill effect. So, when the actual shaft speed is greater than the previous speed threshold, the engine signal is immediately brought to zero; otherwise, it is calculated as the standard procedure illustrated in 4.1.2.

On the contrary, for pitch values lower than φ_0 (i.e. pitch values corresponding to reverse running of the vessel the assessment of pitch and shaft speed set-points is managed in according to a proper combinatory law studied for the crash stop.

4.3 Real Time Hardware In the Loop (RT-HIL)

The procedure developed to design a real propulsion control system can be subdivided into four main steps. The analysis of customer requirements and technical specifica-

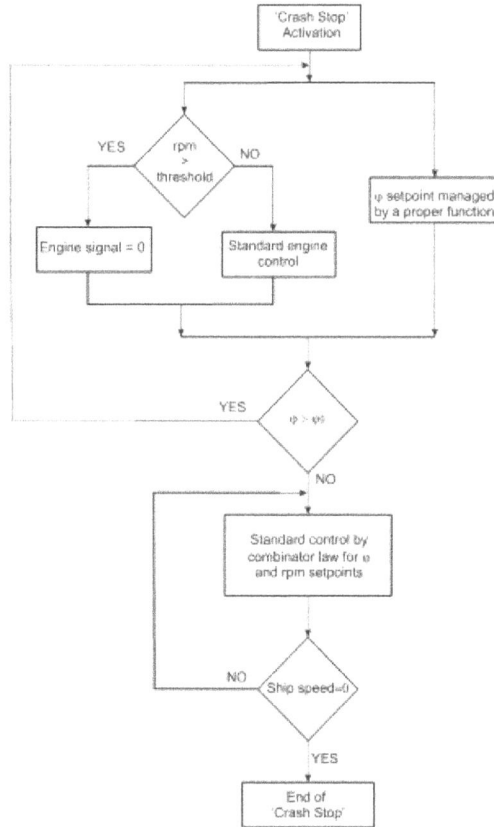

Figure 4.10: Flow chart for the "Crash Stop" control logic

tions leads to a first tentative controller design based on previous experience and expert knowledge of control systems. The second step is a full simulation approach that enables a complete understanding of both steady state and dynamic behaviours of the regulated ship propulsion plant. The third step is the systematic analysis and review of the simulation results, which allows the controller design to be updated and refined until the desired performance is obtained. The implementation of the controller scheme (resulting from the simulations) into the controller software gives the real controller. The fourth step is the debugging and fine tuning of the real controller by a 'Real Time' (RT) 'Hardware In the Loop' (HIL) simulation.

The last step allows the designer to study, in a simulated environment, the performance of the real controller. The procedure integrates basic knowledge and manufacturer experience with massive use of numerical simulation in real time.

During the simulation-based preliminary design, it is appropriate for simulations to run in batch mode to obtain the calculated data more quickly (that means that a

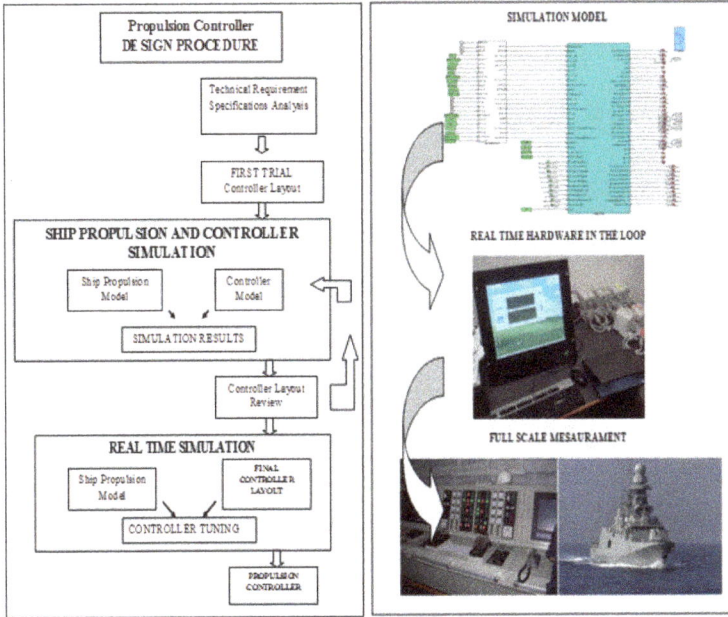

Figure 4.11: Controller design by simulation

60-second test, for example, is simulated by PC in less than 60 seconds). The calculation time depends on several factors regarding both hardware limitations (computer memory, CPU speed) and the developed numerical model (simulation time step, stiffness, kind of solver for the ordinary differential equations). In this phase, the control logic can be designed and several parameters of the regulation loops can be set in an ideal system, representing the interaction between two software (control system and ship) as illustrated in Fig. 4.12.

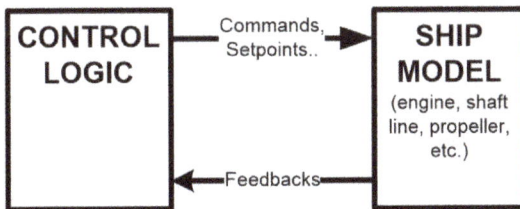

Figure 4.12: Data exchange between control and ship simulator

A typical architecture of the propulsion control system of a ship could be the following:

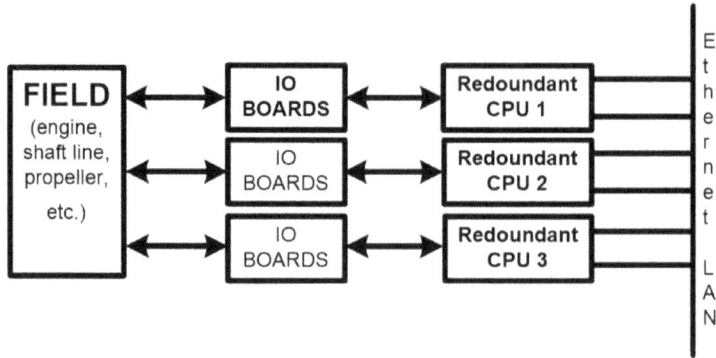

Figure 4.13: Architecture of the propulsion controller

Several CPUs are usually used to control different components of the propulsion system and to limit the loss of functionalities in the case of failure of one of them (even if each CPU is used in a redundant configuration).

Unfortunately, the behaviour of the real hardware on board could be quite different from that of the one simulated during the preliminary design phase.

The main differences could be due to the cyclic time of the CPUs, a different time delay in exchanging data between controllers, and the native functions that can be implemented; further differences could be explained by the presence of many functionalities usually not implemented in the ship numerical model (but that still interact with the propulsion control) and the thousands of signals that the automation has to monitor on the real system.

In the real system, the whole control system has to work in real time and the automation designer has to be sure that the performance foreseen by simulation will be maintained in a real environment. To this end, it is necessary to limit, as soon as possible, most of the differences between the two worlds. This could be possible by the adoption of the RT-HIL method.

Once the control logic performance has been evaluated using batch simulations, the control subsystem is tested by RT-HIL simulation.

This design technique is employed in test setup where the real hardware controller can exchange data with the ship propulsion models (engine, shaft line, propeller, ship motions, etc.) that are simulated in real-time (for example, a 60-second test must take exactly 60 seconds to run on PC).

Generally, the real controller test is performed on board, partially during the delivery period and completely during ship full-scale trials. These trials are time consuming and very expensive as they require the full ship availability. By using RT-HIL simulation, the physical availability of the ship is not required; thus, the controller testing can be done even before the ship is built. In order to increase the simulation realism, some functionalities, not implemented in the ship model, are simulated by codes inside the controllers.

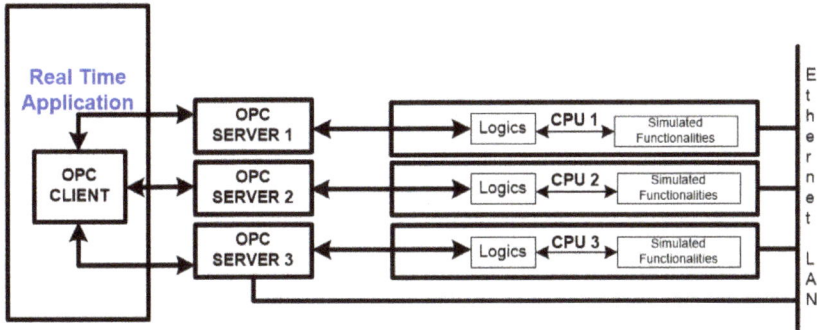

Figure 4.14: Test-bed used for RT HIL trials

Fig. 4.14 shows the block diagram of the experimental setup used for testing the controller. A real-time application executes the ship model. An OPC client reads the command parameters on the controllers through OPC servers and returns the results. OPC servers and the application reside on the same PC and each OPC Server exchanges data with one controller through Ethernet LAN.

The "Simulated Functionalities" can interact with an HMI (Human-Machine Interface) page, seen in Fig. 4.15, to supply commands from users (i.e. the selection of the lever position and of a precise propulsion mode) and to simulate the minor systems not implemented inside the model. All of the parameters exchanged via OPC can be logged to a file, but they are also available to the automation designer by means of a graphical panel as illustrated in Fig. 4.16. In this way, it is possible to obtain a comprehensive view of all system working parameters.

Because the real time application (ship model) is completely configurable, it is possible to try the controller with different working scenarios (i.e. different ambient temperatures, rudder angles, etc.). This is an additional advantage of the RT-HIL approach, not only because the designer can debug the controller before the on-board delivery to maximise the controller performance and minimize costs and delivery time, but also because it is possible to test the system in borderline situations that can be difficult or risky to test with the real ship.

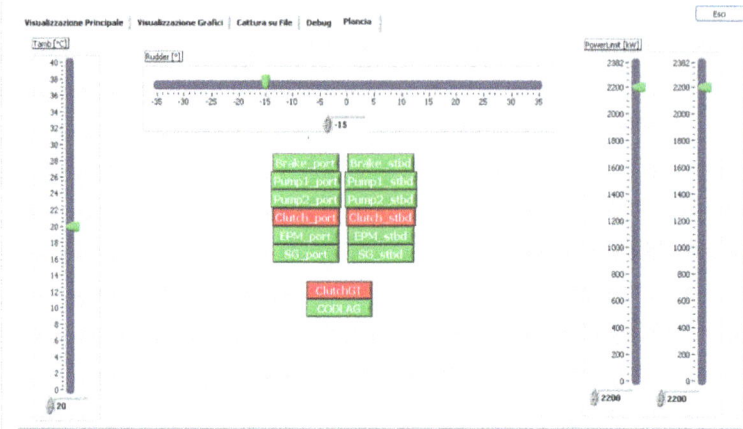

Figure 4.15: Snapshot of simulation input panel

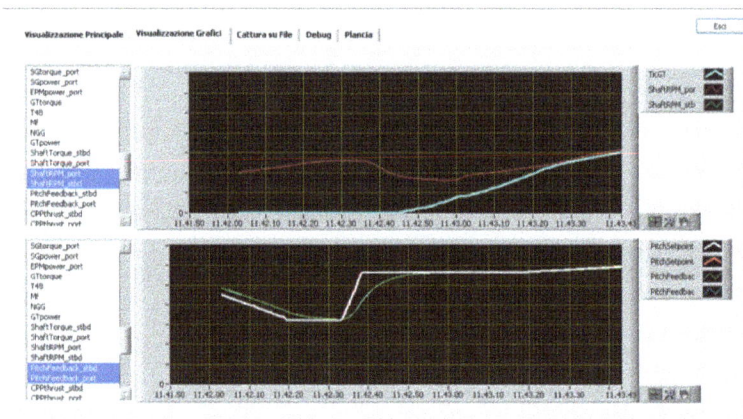

Figure 4.16: Snapshot of simulation output panel

5 Motion Equations

The manoeuvrability problem has been recognized by the international community only in recent years, particularly after the Amoco Cadiz accident in 1978. The ship, due to a storm, grounded in the English Channel and lost its crude oil cargo with catastrophic consequences for the marine environment.

Afterwards, 15 years passed (1993) before the IMO issued minimum standards to ensure the safety of ships, and after another 10 years (2003), these regulations became mandatory.

During the preliminary design stage, it is difficult to evaluate the boat manoeuvreing characteristics without recourse to the model tests; because of this, the numerical simulation becomes an indispensable tool.

Usually, the manoeuvrability problem is treated as a plane problem (with 3 degrees of freedom: surge, sway, and yaw), but our intention is to develop a model able to evaluate all six of the ship motions (thus including heave, roll, and pitch).

By using the same model, it is also possible to study unusual propulsion control system applications such as target tracking, path following, dynamic positioning, and so on.

To develop a unique model suitable for different ships, all the variables in this methodology are expressed in non-dimensional form according to Table 5.1. An important assumption is that all the hydrodynamic coefficients used in the following table are frequency independent.

Table 5.1: Scaling factor

VARIABLES	UNITS (I.S.)	SCALING FACTOR
Linear Position	[m]	L
Linear velocity	[m/s]	V
Angular Velocity	[rad/s]	$\frac{V}{L}$
Mass	[kg]	$\frac{1}{2}\rho L^3$
Mass moment of inertia	[kg m^2]	$\frac{1}{2}\rho L^5$
Force	[N]	$\frac{1}{2}\rho L^2 V^2$
Moments	[Nm]	$\frac{1}{2}\rho L^3 V^2$

5.1 Reference Frame

Before introducing the equations of motion, it is necessary to briefly outline the kinematics of the vessel. Different reference frames are used, so we shall begin our description with the introduction of all frames involved:

– Earth-fixed frame (**n**)

The **n**-frame $(O, \underline{n}_1, \underline{n}_2, \underline{n}_3)$ is a local geographical frame fixed to the Earth. The positive unit vector $\underline{n}_1$ points towards the north, $\underline{n}_2$ points towards the east, and $\underline{n}_3$ points downwards normal to the Earth's surface the vessel moves on. The origin O is located on the mean water free-surface at an appropriate location. The **n**-frame is considered inertial. This is a reasonable assumption because the velocity of marine vehicles is small enough for the forces due to the rotation of the Earth to be negligible. The **n**-frame is used to define the position of the vessel on the Earth and the wind and current direction.

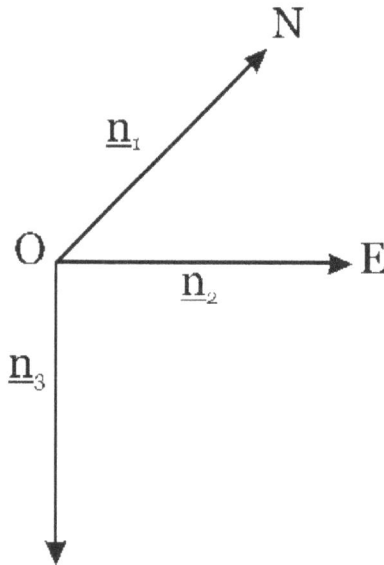

Figure 5.1: Earth Fixed Frame

– Ship-fixed frame (**b**)

The **b**-frame $(O_b, \underline{b}_1, \underline{b}_2, \underline{b}_3)$ is fixed to the vessel hull. The positive unit vector $\underline{b}_1$ points towards the bow, $\underline{b}_2$ points towards starboard and $\underline{b}_3$ points downwards. The centre, O_b, is often located in $\left(\frac{L_{PP}}{2}, 0, z_{WL}\right)$, where L_{PP} is the length between the ahead and the astern perpendicular and z_{WL} is the quota of the water line with respect to the keel line.

The **b**-basis is used to describe velocities, accelerations, and forces, and it is also used to formulate the equations of motion.

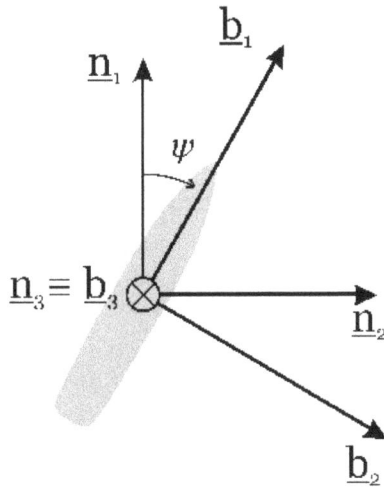

Figure 5.2: Ship Fixed Frame

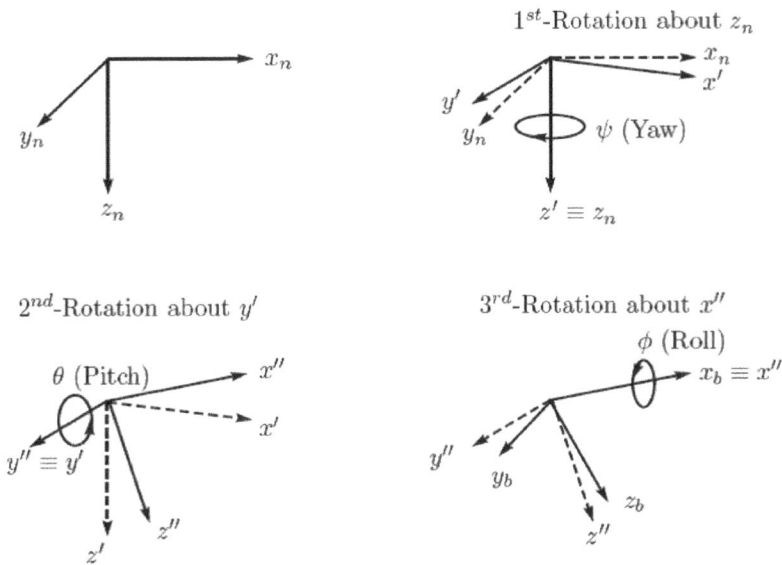

Figure 5.3: Euler Angles

5.2 Kinematics

The position of a vessel is defined by the coordinates of the origin of the **b**-frame relative to the **n**-frame. The attitude (or orientation) of a vessel is defined by the orientation of the **b**-frame with respect to the **n**-frame. This is given by three consecutive rotations about the main axes that take the **n**-frame into the **b**-frame. These rotations can be per-formed in a different order (there are 12 different ways of doing this), and each triplet of rotated angles is called a set of Euler angles. The most commonly used set of Euler angles are yaw, pitch, and roll, which corresponds to the rotations performed in the following order:

1. rotation about the z_n axis of angle ψ (yaw angle),
2. rotation about the y' axis of angle θ (pitch angle),
3. rotation about the x'' axis of angle ϕ (roll angle).

Thus, the position and orientation of a vessel is defined by the coordinates of the origin O_b, relative to the **n**-frame, and by the Euler angles. The generalized position vector (position and orientation) of the vessel in **n**-frame is defined as follows:

$$\eta = \left\{ \begin{array}{c} x\,(t) \\ y\,(t) \\ z\,(t) \\ \phi\,(t) \\ \theta\,(t) \\ \psi\,(t) \end{array} \right\}$$

The generalized velocity vector is the time derivative of the position vector:

$$\dot{\eta} = \left\{ \begin{array}{c} \dot{x}(t) \\ \dot{y}(t) \\ \dot{z}(t) \\ \dot{\phi}(t) \\ \dot{\theta}(t) \\ \dot{\psi}(t) \end{array} \right\}$$

It is convenient to represent the velocities in the **b**-basis.

The usual notation is:

$$v \left\{ \begin{array}{c} u(t) \\ v(t) \\ w(t) \\ p(t) \\ q(t) \\ r(t) \end{array} \right\}$$

Where:

v can be expressed as $v = [\underline{v}^b_{b/n}\ \underline{\omega}^b_{b/n}]$

$\underline{v}^b_{b/n} = \left[\begin{array}{ccc} u, & v, & w \end{array} \right]$ is the linear velocity of the point O_b with respect to the **n**-frame, expressed in the **b**-frame

$\underline{\omega}^b_{b/n} = \left[\begin{array}{ccc} p, & q, & r \end{array} \right]$ is the angular velocity of the **b**-frame with respect to the **n**-frame, expressed in the **b**-basis [41].

Table 5.2: Kinematic Variables

VARIABLE	DESCRIPTION	FRAME/BASIS		
x	North position	n– frame		
y	East position	n– frame		
z	Down position	n– frame		
ϕ	Roll angle	Euler angle n	$\rightarrow$	b
θ	Pitch angle	Euler angle n	$\rightarrow$	b
ψ	Heading or Yaw angle	Euler angle n	$\rightarrow$	b
u	O_b surge velocity	b– basis		
v	O_b sway velocity	b– basis		
w	O_b heave velocity	b– basis		
p	Roll rate	b– basis		
q	Pitch rate	b– basis		
r	Yaw rate	b– basis		

The transportation between the components of the linear velocity, expressed in the **n** -basis and **b** -basis, respectively, is given by the relation:

$$\underline{v}^n_{b/n} = R^n_b \underline{v}^b_{b/n}$$

Where, according to Fig. 5.3, R^n_b is the following matrix [24]:

$$R^n_b = \begin{bmatrix} \cos(\psi)\cos(\theta) & -\sin(\psi)\cos(\phi) + \cos(\psi)\sin(\theta)\sin(\phi) & \sin(\psi)\sin(\phi) + \cos(\psi)\cos(\phi)\sin(\theta) \\ -\sin(\psi)\cos(\theta) & \cos(\psi)\cos(\phi) + \sin(\phi)\sin(\theta)\sin(\psi) & -\cos(\psi)\sin(\phi) + \sin(\theta)\sin(\psi)\cos(\phi) \\ -\sin(\theta) & \cos(\theta)\sin(\phi) & \cos(\theta)\cos(\phi) \end{bmatrix}$$

The following notation is used to transform a generic vector from one basis to another:

$$\underline{v}^{to} = R^{to}_{from}\, \underline{v}^{from}$$

The previous equation means that, to obtain the components v^{to}, it is necessary to apply the matrix R^{to}_{from} on the components v^{from}.

We can evaluate the relationship between the ship angular velocity and its components and the time derivative of the Euler angles through the matrix transportation:

$$\underline{\omega}^n_{b/n} = T^n_b \underline{\omega}^b_{b/n}$$

Where T^n_b is given by:

$$T^n_b = \begin{bmatrix} 1 & \sin(\phi)\tan(\theta) & \cos(\phi)\tan(\theta) \\ 0 & \cos(\phi) & -\sin(\phi) \\ 0 & \dfrac{\sin(\phi)}{\cos(\theta)} & \dfrac{\cos(\phi)}{\cos(\theta)} \end{bmatrix}$$

It is important to note that the matrix has a singularity when the pitch angle $\theta = \pm\frac{\pi}{2}$; however, for our aim, this does not create any problems because the ship will be never operate in this condition.

In summary, we can write the six degrees of freedom kinematic equations in a compact way as [25]:

$$\dot{\eta} = \begin{bmatrix} R_b^n & 0_{3x3} \\ 0_{3x3} & T_b^n \end{bmatrix} v$$

5.3 6 D.O.F. Motion Equation

The horizontal motion includes only 3 significant degrees of freedom: x (surge), y (sway), z (yaw).

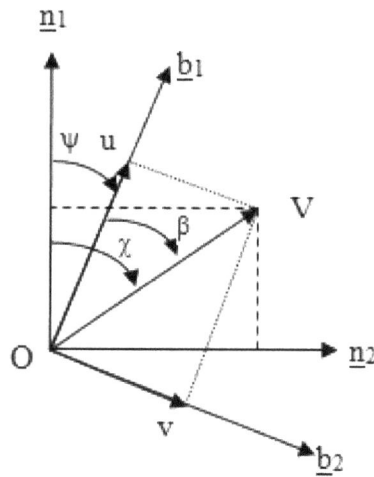

Figure 5.4: Angle Notation

In accordance with Fig. 5.4, some definitions are given:
- Heading angle ψ defines the direction of the vessel bow; it is positive by the right hand screw convention.
- Course angle χ defines the direction of the velocity vector of the vessel; it is positive by the right hand screw convention.
- Drift angle β defines the direction of the velocity vector of the vessel with respect to the bow; it is positive by the right hand screw convention.

The relationship between the three previous angles is given by:

$$\chi = \psi + \beta$$

It is also possible to evaluate the drift angle as a function of the velocity vector components:

$$\beta = \text{atan}\left(\frac{v}{u}\right)$$

A generic rigid body motion requires 6 generalized coordinates (3 displacements and 3 rotations) to be identified. When all six coupled motions (surge, sway, heave, roll, pitch, and yaw) are concerned, a vessel can be considered as a six degrees-of-freedom (d.o.f.) rigid-body as represented in Fig. 5.5.

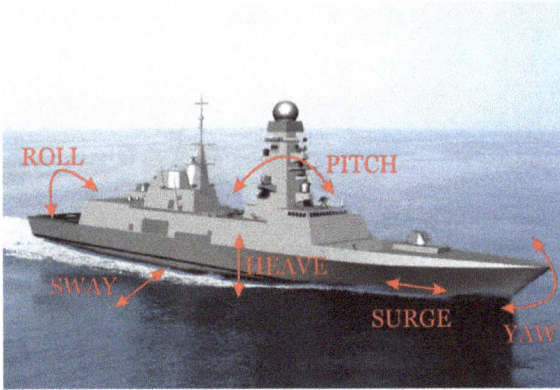

Figure 5.5: Six degrees of freedom

The corresponding equations of motion can be deduced by the forces and moments equations using the second Newton's law:

$$m \cdot \dot{v}_G = \tau$$

Where:
  $\dot{v}_G$ is the ship gravity centre acceleration with respect to the inertial frame **n**
  Due to the rotation of the **b**-frame with respect to **n**-frame, is it necessary to introduce further terms that identify the Coriolis and centripetal forces and moments. Then, to develop the motion equations for an arbitrary origin (different from the gravity centre) the following equation [25] has been used:

$$\tau = M_{RB}\dot{v} + C_{RB}(v)v$$

Where:
  M_{RB} is the inertia matrix
  C_{RB} is the inertia Coriolis-centripetal matrix

τ are the total forces and total moments acting on the vessel

$$M_{RB} = \begin{bmatrix} m & 0 & 0 & 0 & mz_G & -my_G \\ 0 & m & 0 & -mz_G & 0 & mx_G \\ 0 & 0 & m & my_G & -mx_G & 0 \\ 0 & -mz_G & my_G & I_{xx} & I_{xy} & I_{xz} \\ mz_G & 0 & -mx_G & I_{yx} & I_{yy} & I_{yz} \\ -myz_G & mx_G & 0 & I_{zx} & I_{zy} & I_{zz} \end{bmatrix}$$

Due to the ship symmetry: $I_{xy} = I_{yx} = I_{yz} = I_{zy} = 0$ and $y_G = 0$.

The Coriolis-centripetal matrix can be represented in different ways, but for our aim we choose the parameterization [39] as a function of the inertia matrix:

$$C_{RB}(v) = \begin{bmatrix} 0_{3x3} & -S\left(M_{11}v^b_{b/n} + M_{12}\omega^b_{b/n}\right) \\ -S\left(M_{11}v^b_{b/n} + M_{12}\omega^b_{b/n}\right) & -S\left(M_{21}v^b_{b/n} + M_{22}\omega^b_{b/n}\right) \end{bmatrix}$$

Where:

S denotes the cross-product operator:

$$\underline{a} \wedge \underline{b} = S(\underline{a})\,\underline{b} = \begin{bmatrix} 0 & -a_3 & a_1 \\ a_3 & 0 & -a_2 \\ -a_1 & a_2 & 0 \end{bmatrix} \begin{bmatrix} b_1 \\ b_2 \\ b_3 \end{bmatrix}$$

M_{ij} is the 3×3 sub-block of the matrix M_{RB}

Coupling the kinematic equation and Newton's law, we obtain a system of twelve first order differential equations whose solutions provide the instantaneous position of the gravity centre (ship trajectory) and the velocity vector.

The main difficulties for solving the differential equation system are the correct evaluations of the forces acting on the ship.

5.4 Proposed Forces Evaluation

It has been hypothesized that the forces and the moments acting on the vessel are the following:
-   Hull forces and moments τ_H
-   Propeller forces and moments τ_P
-   Rudder forces and moments τ_R
-   Environmental forces and moments τ_W
-   Restoring Forces and moments τ_G

So the total external forces acting on a ship can be represented by the sum:

$$\tau = \tau_H + \tau_P + \tau_R + \tau_W + \tau_G$$

5.4.1 Hull Forces

The hull force contribution can be decomposed into three main components: added mass effect τ_A, linear τ_{LD}, and non-linear damping effects τ_{NLD}:

$$\tau_H = \tau_A + \tau_L + \tau_{NL}$$

The added mass forces are the hydrodynamic forces when a body is forced to accelerate, but these terms are not to be considered like a mass increase of the vessel.

Regarding the added mass forces, it is possible again to decompose them into two components:

$$\tau_A = M_A \dot{v} + C_A(v)v$$

Where:

M_A is the added mass matrix

C_A is the hydrodynamic Coriolis-centripetal matrix

The coefficients of the matrix M_A can be derived in different ways, such as statistical data available in literature, strip theory, potential theory, CFD techniques, and experimental trial.

The proposed added mass matrix is obtained starting from statistical data, and it is a particular case of the problem. Not all the components are expressed due to the fact that some data were not available:

$$M_A = \begin{bmatrix} X_{\dot{u}} & 0 & 0 & 0 & 0 & 0 \\ 0 & Y_{\dot{v}} & 0 & Y_{\dot{p}} & 0 & 0 \\ 0 & 0 & Z_{\dot{w}} & 0 & 0 & 0 \\ 0 & K_{\dot{v}} & 0 & K_{\dot{p}} & 0 & K_{\dot{r}} \\ 0 & 0 & 0 & 0 & M_{\dot{q}} & 0 \\ 0 & 0 & 0 & N_{\dot{p}} & 0 & N_{\dot{r}} \end{bmatrix}$$

The notation [41] is also used to define added mass matrix and damping matrix components.

I.e. $X_{\dot{u}}$ denotes the coefficient of the added mass force along the $\underline{b}_1$ direction due to an acceleration $\dot{u}$ along the $\underline{b}_1$ direction.

The C_A can be easily calculated in the same way of the matrix C_{RB}, but in our case, we did not consider this term because it is included in the damping coefficient matrix due to the fact that the usual way to obtain hydrodynamic coefficients by means of model tests (or regression formulas derived by them) implicitly includes their effects.

$$C_A = 0_{6x6}$$

Damping forces are given as a function of the velocity vector, so the contribution of damping effects can be calculated through the following formula:

$$\tau_D = D_L v + D_{NL}(v)v$$

Where:

D_L is the linear damping matrix

D_{NL} is the non-linear damping matrix

$$D_L = \begin{bmatrix} X_u & 0 & 0 & 0 & 0 & 0 \\ 0 & Y_v & 0 & Y_p & 0 & Y_r \\ 0 & 0 & Z_w & 0 & 0 & 0 \\ 0 & K_v & 0 & K_p & 0 & K_r \\ 0 & 0 & 0 & 0 & M_q & 0 \\ 0 & N_v & 0 & N_p & 0 & N_r \end{bmatrix}$$

I.e. Y_v denotes the coefficient of the linear damping force along $\underline{b}_2$ direction due to a velocity v along the $\underline{b}_2$ direction.

$$D_{NL} = \begin{bmatrix} 0 & X_{vv}v + X_{vr}r & 0 & 0 & 0 & X_{rr}r \\ 0 & Y_{v|\varphi|}\,|\varphi| + Y_{vrr}r^2 + Y_{v|v|}\,|v| + Y_{v|v||\phi|}\,|v|\,|\varphi| & 0 & Y_{p|p|}\,|p| & 0 & Y_{r|\varphi|}\,|\varphi| + Y_{r|r|}\,|r| + Y_{r|v|}\,|v| + Y_{r|r||\phi|}\,|r|\,|\varphi| \\ 0 & 0 & Z_{w|w|}\,|w| & 0 & 0 & 0 \\ 0 & K_{v|\varphi|}\,|\varphi| + K_{v|v|}\,|v| + K_{vrr}r^2 + K_{v|v||\varphi|}\,|v|\,|\varphi| & 0 & K_{p|p|}\,|p| & 0 & K_{r|\varphi|}\,|\varphi| + K_{r|r|}\,|r| + K_{r|v|}\,|v| + K_{r|r||\varphi|}\,|r|\,|\varphi| \\ 0 & 0 & 0 & 0 & M_{q|q|}\,|q| & 0 \\ 0 & N_{v|\varphi|}\,|\varphi| + N_{v|v|}\,|v| + N_{vrr}r^2 + N_{v|v||\varphi|}\,|v|\,|\varphi| & 0 & N_{p|p|}\,|p| & 0 & N_{r|\varphi|}\,|\varphi| + N_{r|r|}\,|r| + N_{r|v|}\,|v| + N_{r|r||\varphi|}\,|r|\,|\varphi| \end{bmatrix}$$

The elements of D_L and D_{NL} matrix are evaluated using regression [30, 43] corrected first to take into account appendage and then corrected through a model scale experimental campaign.

5.4.2 Propulsion Forces

The propulsion forces and moments τ_P belong to this force class. In this case, the propulsion forces and moments are generated by two controllable pitch propellers. They include: longitudinal force (thrust) influenced by asymmetrical functioning during manoeuvres (via asymmetrical propulsive coefficients); the lateral and vertical forces due to the oblique propeller flow. The procedure to evaluate these kinds of forces are explained in Paragraph 6.4.2.

5.4.3 Rudder Forces

They are the forces and the moments developed by the rudder τ_R. In this case, the steering forces and moments are generated by two spade rudders, which can also have a rudder roll functioning (roll stabilization in adverse environment). The procedure to evaluate these kinds of forces is explained in 7.3.

5.4.4 Environmental Forces

There are three types of environmental disturbances acting on the ship: wind, current, and wave. In this analysis, however, none of these effects was taken into account.

5.4.5 Restoring Forces

The restoring forces and moments τ_G are due to buoyancy $B_s = \rho g \nabla$ and weight force $W_s = mg$, where m is the mass of the vessel, g is the gravity acceleration, ρ is the sea water mass density, and ∇ is the fluid displaced volume.

The analytical expression of τ_G is derived on the basis of geometrical and inertial properties of the vessel. This kind of force is position-dependent and is strongly influenced by ship load condition.

The restoring forces can be written in a compact vectorial form:

$$\tau_G = G\eta$$

Where the G matrix is shown below; its components are calculated according to [34]:

$$G = \begin{bmatrix} 0 & 0 & 0 & 0 & 0 & 0 \\ 0 & 0 & 0 & 0 & 0 & 0 \\ 0 & 0 & \rho g A_w & 0 & -\rho g A_w x_f & 0 \\ 0 & 0 & 0 & gmGMT & 0 & 0 \\ 0 & 0 & -\rho g A_w x_f & 0 & gmGML & 0 \\ 0 & 0 & 0 & 0 & 0 & 0 \end{bmatrix}$$

Where:
  A_w is the water plane area
  x_f is the longitudinal coordinate of the floatation centre
  GMT is the transversal metacentric height
  GML is the longitudinal metacentric height
  The restoring matrix evaluated with the previous formulation is linear, and for a more precise formulation it would be necessary to use the righting arm GZ instead of GMT. Nevertheless, the small angles usually involved in the manoeuvrability problem allow for this simplification.

Summarizing the results, the motion equations can be written in the following compact form:

$$(M_{RB} + M_A)\dot{v} = C_{RB}(v)v + D_L v + D_{NL}(v)v + G(\eta)\eta + \tau_P + \tau_R$$

6 Propeller & Pitch Change Mechanism

One of the major targets for the propeller design is a high propeller efficiency, but is also important to optimize the propeller for noise, vibration, and cavitation.

A controllable pitch propeller (CPP) is indeed more complex than a fixed propeller; it contains more parts, many of them moving, and it would be unrealistic to expect that there is no extra risk entailed in installing CPPs rather than fixed propellers.

Unlike fixed pitch propellers, whose only operational variable is rotational speed, the controllable pitch propeller provides an extra degree of freedom in its ability to change blade pitch. However, for some propulsion applications, particularly those involving shaft-driven generators, the shaft speed is held constant. This reduces the number of operating variables, again, to one. It is important to evaluate a correct relationship between shaft speed and pitch angle. This relationship is called a combinatory curve.

Using the combinatory curve, the efficiency is maximised and the risk of cavitation is minimized.

In the last forty years, the controllable pitch propeller has grown in popularity from representing a small proportion of the propellers produced to its current very substantial market share.

This growth is illustrated in Fig. 6.1, which shows the proportion of controllable pitch propeller systems when compared to the total number of propulsion systems classified in Lloyd's Register during the years 1960 to 2004, taken at five-year intervals.

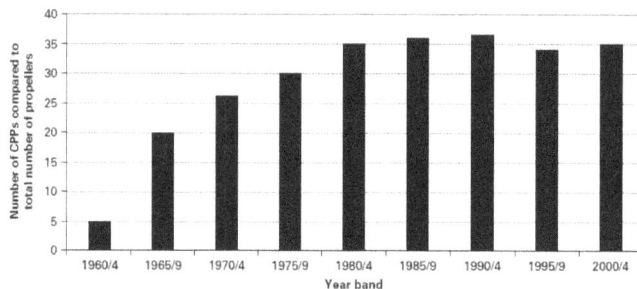

Figure 6.1: Market share of controllable pitch propellers

Fig. 6.1 shows the relative distribution of controllable pitch propellers within certain classes of ship type.

From Table 6.1, the controllable pitch propeller is currently most favored in vessels with several operative speed profiles such as passenger ships and ferries, general cargo ships, and tugs and fishing vessels.

Table 6.1: Percentage of CPP compared to the total number of propellers by ship type

Ship type	1960–1964	1965–1969	1970–1974	1975–1979	1980–1984	1985–1989	1990–1994	1995–1999	2000–2004
Tankers	1	7	15	14	23	13	21	17	10
Bulk carriers	1	9	10	5	5	12	0	1	1
Container ships	0	13	24	3	1	13	18	10	9
General cargo	2	12	20	29	42	43	45	55	80
Passenger ships and ferries	24	64	82	100	94	100	88	78	63
Tugs and offshore vessels	29	50	44	76	85	100	77	73	78
Fishing vessels	48	54	87	90	93	92	100	90	89

6.1 Pitch Change Mechanism

Controllable pitch propellers are generally actuated by hydraulic oil power systems due to their high power/volume ratios. Usually, the simplest hydraulic-mechanical actuator is used: the piston in a cylinder. The oil flows from the tank to the OD box in a normal pipe inside the engine room. The OD box is a directional valve located on the shaft from which oil flows through a twin pipe inside the shafting to the propeller hub piston. Inside the CPP hub, a double effect hydraulic cylinder is actuated by the oil pressure in a longitudinal direction. Two actuating chambers are needed to move the blade at both positive and negative pitch angles.

The pitch can be measured with the feedback pipe, a potentiometer attached to the inner transmission line. For a more accurate pitch measurement, the sensors can be positioned inside the hub. The chosen pitch angle is achieved by the proportional pitch controller that compares the actual pitch set-point with the real pitch and manipulates the oil valve.

The system layout is schematized in Fig. 6.2.

The propeller blades are connected to the hub with different kinds of suspension.

Usually, the blades are connected to the hydraulic actuator in the hub via a slot-pin mechanism.

For the pin-slot mechanism, the relationship between the hydraulic force generated inside the hub (piston force) and the spindle torque is the following [46]:

$$Q_{hyd} = \frac{F_{hyd} e_{yp}}{\cos^2(\varphi)}$$

Where:

Q_{hyd} hydraulic torque
F_{hyd} [N] hydraulic force
e_{yp} eccentricity of the yoke pin
φ pitch angle

The relationship between the linear movement (piston stroke) and the pitch angle (blade angular position) is expressed as:

$$x_{pist} = x_{MAX\_AH} - \tan(\varphi_{MAX\_AH} - \varphi) e_{yp}$$

Where:

x_{pist} instantaneous linear position (stroke)

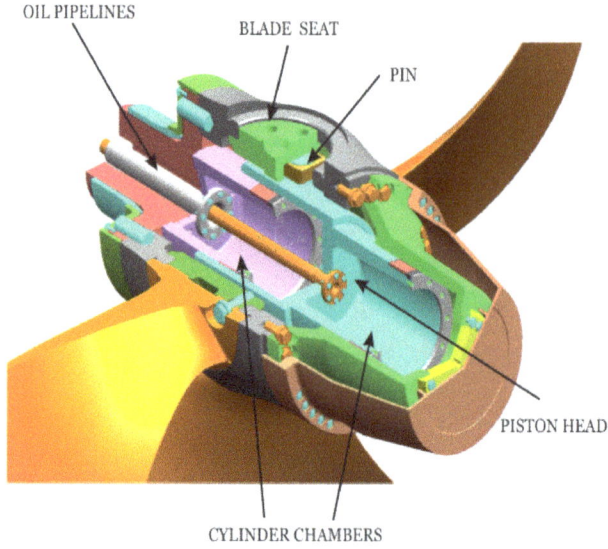

Figure 6.2: Main elements inside a CPP hub.

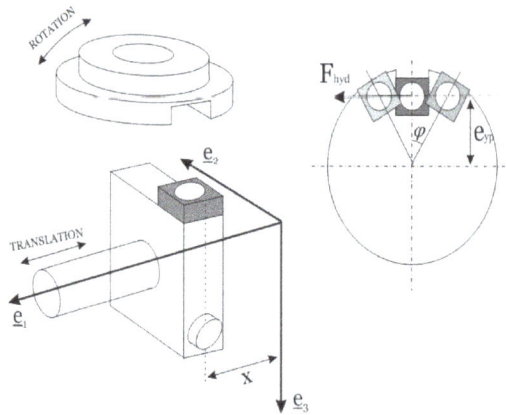

Figure 6.3: Pin-slot mechanism

x_{MAX_AH} maximum stroke ahead

φ_{MAX_AH} maximum pitch ahead

All the variables appearing on the right hand side of the relationship between the linear movement and the pitch angle are derived from the corresponding mechanism drawing represented in Fig. 6.4.

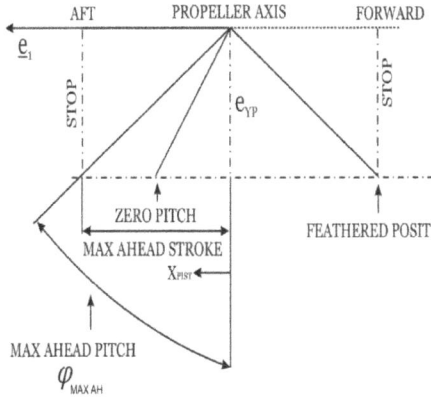

Figure 6.4: Relationship between linear and angular position

6.2 Reference Frame

Before introducing the equations for blade motion and the methodology for evaluation of the forces, it is necessary to briefly outline the reference frames. Different reference frames are used, so it is best to define the coordinate system involved:

– Hub-fixed frame (**e**)

The **e**-frame $\left(O_e, \underline{e}_1, \underline{e}_2, \underline{e}_3\right)$ is fixed to the hub. The positive unit vector $\underline{e}_1$ points towards the bow, $\underline{e}_3$ coincides with the unit vector of the spindle axis, and $\underline{e}_2$ is the vectorial product of $\underline{e}_3$ and $\underline{e}_1$. The origin O_e of the frame is in the centre of the shaft line.

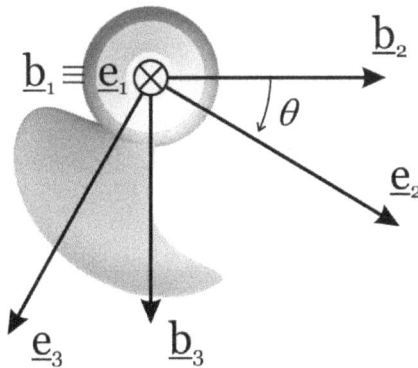

Figure 6.5: Hub Fixed Frame **e**

The relationship between **e**-frame and **b**-frame is defined as the following:

$$
\begin{cases}
\underline{e}_1 = \underline{b}_1 \\
\underline{e}_2 = \cos\vartheta \underline{b}_2 + \sin\vartheta\, \underline{b}_3 \\
\underline{e}_3 = -\sin\vartheta \underline{b}_2 + \cos\vartheta \underline{b}_3
\end{cases}
\qquad
\begin{cases}
\underline{b}_1 = \underline{e}_1 \\
\underline{b}_2 = \cos\vartheta\, \underline{e}_2 - \sin\vartheta\, \underline{e}_3 \\
\underline{b}_3 = \sin\vartheta \underline{e}_2 + \cos\vartheta \underline{e}_3
\end{cases}
$$

– Blade-fixed frame (**f**)

The **f**-frame $\left(O_f, \underline{f}_1, \underline{f}_2, \underline{f}_3\right)$ is fixed to the blade. The positive unit vectors $\underline{f}_1$ and $\underline{f}_2$ are the unit vectors $\underline{e}_1$ and $\underline{e}_2$ rotated by a pitch angle of the blade φ, and $\underline{f}_3$ coincides with $\underline{e}_3$. The origin O_f of the frame is in the centre of the shaft line with respect to the vertical position and under the spindle axis for the witch regarding the longitudinal and transversal position.

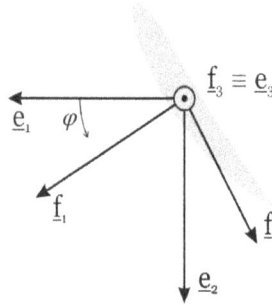

Figure 6.6: Blade Fixed Frame (**f**)

The relationship between **f**-frame and **e**-frame is shown by following:

$$
\begin{cases}
\underline{f}_1 = \cos\varphi \underline{e}_1 + \sin\varphi\, \underline{e}_2 \\
\underline{f}_2 = -\sin\varphi \underline{e}_1 + \cos\varphi\, \underline{e}_2 \\
\underline{f}_3 = \underline{e}_3
\end{cases}
\qquad
\begin{cases}
\underline{e}_1 = \cos\varphi \underline{f}_1 - \sin\varphi \underline{f}_2 \\
\underline{e}_2 = \sin\varphi\, \underline{f}_2 + \cos\varphi \underline{f}_3 \\
\underline{e}_3 = \underline{f}_3
\end{cases}
$$

Using the angular velocity composition theorem, it is possible to represent the angular velocities (with respect to the inertial **n**-frame) of a single blade $\underline{\omega}_B$ and the hub $\underline{\omega}_H$ respectively, as:

$$
\underline{\omega}_B = \underline{\omega}_\psi + \underline{\omega}_\vartheta + \underline{\omega}_\varphi = \dot{\psi}\underline{b}_3 + \dot{\vartheta}\underline{e}_1 + \dot{\varphi}\underline{f}_3
$$

$$
\underline{\omega}_H = \underline{\omega}_\psi + \underline{\omega}_\vartheta = \dot{\psi}\underline{b}_3 + \dot{\vartheta}\underline{e}_1
$$

where $\dot{\psi}$, $\dot{\vartheta}$ and $\dot{\varphi}$ [rad/s] denote the time derivatives of the yaw angle, the blade angular position, and the pitch angle, respectively.

6.3 Motion and Pressure Differential Equation

The blade position is the key factor for the ship performance, whereas the oil pressure is the key parameter of the mechanism; excessively high pressures can be responsible for mechanism failure, and hence they should be always monitored by the ship automation systems. Thus the proposed mathematical model, developed by the author [9], is based on two differential equations describing the motion of the blade and the motion of the piston, respectively.

The first differential equation is the equation of motion of a blade around its $\underline{f}_3$ axis evaluated in the **e**-frame:

$$\ddot{\varphi}(t) = \frac{1}{I_b} \left(Q_{hyd}(t) + Q_S(t) + Q_{-\Phi}(t) \right)$$

Where:

$\ddot{\varphi}$ is the blade angular acceleration

$Q_{-\Phi}$ is the torque due to the interaction forces between propeller blade and blade bearing

Q_{hyd} is the hydraulic torque

Q_s is the total spindle torque acting on the blade

I_b is the moment of inertia of the blade about the spindle axis $\underline{f}_3$

The second differential equation describes the motion of the cylinder:

$$m_m \ddot{x}_{pist}(t) = A_1 p_1(t) - A_2 p_2(t) - B_p \dot{x}_{pist}(t) + \sum_{i=1}^{z} \Phi_i(t)$$

Where:

$m_m = m_y + m_p + m_o$ is the sum of the yoke mass (m_y), propeller mass (m_p), and oil mass (m_o)

$\ddot{x}_{pist}$ is the piston acceleration

A_1 and A_2 are the yoke areas of the astern chamber and of the ahead chamber, respectively

p_1 and p_p are the pressures inside the two chambers

B_p is the damping coefficient

$\sum_{i=1}^{z} \Phi_i$ is the resultant of the reaction forces due to the interaction between each blade and the bearing

Z is the number of blades

In order to properly implement the differential equations, all the forces and moments acting on a single blade have to be evaluated. This will be shown in the following sections.

6.4 Forces Evaluation

In order to develop a simulation model suitable for CPP control system design and testing, current knowledge of the loads acting on a blade must be assessed.

The total spindle torque is the torque acting around the spindle axis of the propeller blade, and it must either be balanced by the hub mechanism to hold the blades in the required pitch setting or, alternatively, be overcome when a pitch change is required. The spindle torque can be subdivided into three components as follows [19]:

$$Q_S(J, \varphi) = Q_{SH}(J, \varphi) + Q_{SI}(J, \varphi) + Q_{SF}(J, \varphi)$$

Where:

Q_S is the total spindle torque at a given value of non-dimensional propeller advance coefficient J and at a given value of pitch angle φ;

Q_{SH} is the hydrodynamic component of spindle torque due to the water pressure field acting on the blade surfaces;

Q_{SI} is the inertial component (Q_{SC} in Carlton nomenclature) resulting from the blade mass distribution;

Q_{SF} is the frictional component of spindle torque resulting from the relative motion of the surfaces within the blade hub.

The spindle torques are usually expressed in the non-dimensional forms of KQ_{SC} and KQ_{SI}. These coefficients are similar to the conventional propeller torque coefficient, and they are related to the respective spindle torques components as follows:

$$KQ_{SH} = \frac{Q_{SH}}{\rho n^2 D^5}$$

$$KQ_{SI} = \frac{Q_{SI}}{\rho_b n^2 D^5}$$

Where:

ρ and ρ_b are the water and the blade mass densities, respectively

n is the propeller speed

D is the propeller diameter

Little data about spindle torque has been published. Some of the most exhaustive work is the experimental data of the 3-BLADED JP-CPP SERIES [20]. Unfortunately, this data is not sufficient to accurately represent the behaviour of the CPP actuating mechanism during ship manoeuvres. Some recent publications [12, 29, 44] concern the evaluation of the spindle torque acting on a blade.

First of all, it is necessary to list the forces acting on the controllable pitch propeller blades:

– Inertial Forces τ_{PI}
– Hydrodynamic Forces τ_P
– Frictional Forces τ_{PF}
– Hydraulic Forces F_{hyd}

So the total external forces acting on a blade are:

$$\tau_{PROP} = \tau_{PI} + \tau_P + \tau_{PF} + F_{hyd}$$

6.4.1 Inertial Forces

The motion of the ship and the rotation of the propeller give rise to corresponding Coriolis and transportation inertial forces acting on each blade in the **e**-frame. Moreover, gravity yields a sinusoidal varying force.

In more detail, the Coriolis force is defined by:

$$\underline{F}^C = \int_\xi -2\rho_b \underline{\omega}_H \wedge \underline{v}_P^r d\tau$$

The transportation force is expressed as:

$$\underline{F}^S = \int_\xi -\rho_b \left[\underline{a}_o + \underline{\omega}_H \wedge \left(\underline{\omega}_H \wedge (P - O) \right) + \underline{\dot{\omega}}_H \wedge (P - O) \right] d\tau$$

The weight force is given by:

$$\underline{F}^W = \int_\xi \rho_b \underline{g} d\tau = m_b \underline{g}$$

Where:

$\underline{a}_o$ is the linear acceleration of the origin O with respect to the inertial frame

$\rho_b g$ is the mass density of the propeller blade

$(P - O)$ is the position vector of a generic point P of the blade with respect to the origin O

m_b is the blade mass

g is the gravity acceleration

ξg is the whole set of points constituting the blade

$\underline{v}_P^r$ is the linear velocity vector of a generic point P of the blade evaluated in the **e**-frame

More explicitly:

$$\underline{v}_P^r = \underline{v}_o^r + \underline{\omega}_\varphi \wedge (P - O) = \dot{\varphi} \underline{f}_3 \wedge (P - O)$$

Recalling the definition of centre of gravity for a blade:

$$(G - O) = \frac{1}{m_b} \int_\xi \rho_b \, (P - O) \, d\tau$$

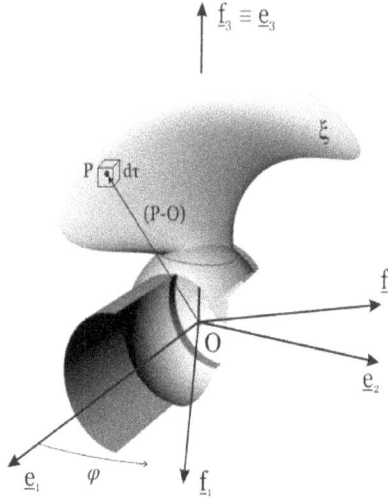

Figure 6.7: Hub and blade reference frames

It is therefore possible to express the inertial forces in a simpler form:

$$\underline{F}^C = -2m_b\underline{\omega}_H \wedge \underline{v}_G^r$$

$$\underline{F}^S = -m_b\underline{a}_o - m_b\underline{\omega}_H \wedge \left[\underline{\omega}_H \wedge (G - O)\right] - m_b\underline{\dot{\omega}}_H \wedge (G - O)$$

The total contribution of inertial and weight forces acting on a blade is then given by the sum:

$$\underline{F}_I = \underline{F}^W + \underline{F}^S + \underline{F}^C$$

In addition, the moments with respect to the origin O of the above forces have to be evaluated. To this end, the moment of a generic (density of) force acting on a blade is given by:

$$\underline{M}_O = \int_\xi (P - O) \wedge \underline{F} d\tau$$

Combining the equation that provides a generic moment and the definition of the inertia tensor with respect to the origin O for each blade:

$$I_o(\underline{v}) = \int_\xi \rho_b (P - O) \wedge [\underline{v} \wedge (P - O)] \, d\tau$$

It is possible to demonstrate (see Appendix 1 for the complete proof) that the moments of the inertial forces can be expressed as:

$$\underline{M}_O^C = \underline{\omega}_\varphi \wedge I_o(\underline{\omega}_H) - \underline{\omega}_H \wedge I_o\left(\underline{\omega}_\varphi\right) + I_o\left(\underline{\omega}_\varphi \wedge \underline{\omega}_H\right)$$

$$\underline{M}_O^S = -m_b\,(G - O) \wedge \underline{a}_O - \underline{\omega}_H \wedge I_o\,(\underline{\omega}_H) - I_o\,(\underline{\dot{\omega}}_H)$$

Where $\underline{\omega}_\varphi = \dot\varphi \underline{f}_3$ denotes the angular velocity of the blade evaluated in the **e**-frame. The moment due to the weight force is:

$$\underline{M}_O^W = m_b(G - O) \wedge \underline{g}$$

The total moment with respect to the origin O due to inertial and weight forces acting on a blade is then given by the sum:

$$\underline{M}_O^I = \underline{M}_O^W + \underline{M}_O^S + \underline{M}_O^C$$

All moments are evaluated in [Nm]. The component of the vector $\underline{M}_O^I$ along the unit vector $\underline{f}_3$ provides the scalar quantity Q_{SI}.

Composing the vectors $\underline{F}_I$ and $\underline{M}_O^I$, it's possible to obtain τ_{PI}.

6.4.2 Propeller Hydrodynamic Forces

In the literature, different numerical methods based, for instance, on the potential approach or on R.A.N.S.E. solver [27, 28], have been proposed to predict propeller hydrodynamic loads. Unfortunately, due to their long computational times, these methods are difficult to apply in the context of a time domain simulator like the one described in the present work. Therefore, in the proposed work, the hydrodynamic forces have been evaluated through a quasi-steady-state methodology based on the propeller open water tests. These tests provide an open water diagram, which allows the evaluation of the thrust coefficient K_T and torque coefficient K_Q. For controllable pitch propellers, these coefficients depend on the advance coefficient J and from blade position φ:

$$K_T = \frac{T}{\rho n^2 D^4}$$

$$K_Q = \frac{Q}{\rho n^2 D^5}$$

$$J = \frac{V_a}{nD}$$

By definition, it is possible to calculate the hydrodynamic force along $\underline{b}_1$ direction, the thrust T, and the required propeller torque Q_P using the following formulas:

$$T\,(J,\varphi) = K_T\,(J,\varphi)\rho n^2 D^4$$

$$Q_P\,(J,\varphi) = K_Q\,(J,\varphi)\rho n^2 D^5$$

Where:
   ρ is the sea water density
   D is the propeller diameter

n is the shaft line revolution regime

A clarification is needed to achieve a correct J evaluation. The fluid velocity evaluated in O_f is different from the fluid velocity evaluated in O_b so the rigid motion formula was used to calculate the local (propeller) fluid velocity:

$$\underline{v}^b_{prop}\,(1:3) = \underline{v}^b_{stern}\,(1:3) + \underline{v}^b_{stern}(4:6) \wedge \left(O_b - O_f\right)$$

The first component of $\underline{v}^b_{prop}$ is the effective longitudinal velocity on the propeller blade. Another important aspect to be taken into account is the study of the propeller behaviour during manoeuvring. Twin screw ships may experience considerably asymmetrical propeller function during manoeuvres. This phenomenon may result in large power fluctuations during tight manoeuvres, with increases of shaft torque up to and over 100% of the steady values in straight course and considerable unbalances between the two shaft lines. This phenomenon has been modelled for imbalances of both of wake field and propeller thrust. To take into account the different wake field, on the inner and outer propeller, it is assumed that, during manoeuvres:

$$(1 - w)_{unbal} = (1 - w)_{Straight} + \Delta w$$

Thus:

$$J_{eff} = \frac{\underline{v}^b_{prop}\,(1)\left[(1 - w)_{Straight} + \Delta w\right]}{nD}$$

Where Δw is a correction obtained from the analysis of a set of dedicated experimental trials [21] as a function of the drift angle and ship speed. This approach is schematized in Fig. 6.8; in particular, two effects are superimposed during manoeuvres. The first is a symmetrical variation of the advance coefficient due to speed reduction during turns. The second is an asymmetrical variation of the advance coefficient, which results in asymmetrical loading of shaft lines.

To take into account the thrust unbalances experienced during manoeuvres with respect to open water characteristics, a corrective factor is introduced [42] that is a function of the drift angle and ship speed:

$$\tilde{t} = \frac{T_{unbal}}{T_{openwater}}$$

Where T_{unbal} is the measured thrust from experimental trials, and $T_{openwater}$ is the thrust that would be obtained from the open water diagrams using J_{eff} as the advance coefficient.

In summary, the procedure to evaluate the thrust and the required torque requires the evaluation of the advance coefficient J_{eff} followed by the evaluation of the pitch angle φ. Then, through a surface interpolation, it is possible to evaluate the propeller hydrodynamic coefficients K_T and K_Q as it is shown in Fig. 6.9. Finally, the thrust magnitude is corrected trough the $\tilde{t}$ coefficient. This procedure is efficient from the point of view of computation time; however, the transient precision is difficult to assess.

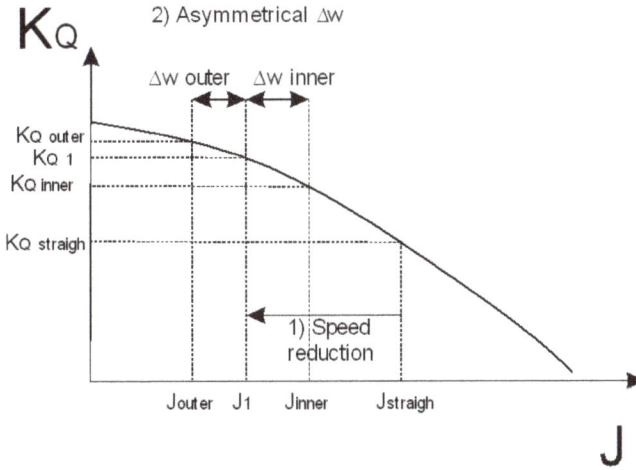

Figure 6.8: Asymmetrical variation of advance coefficient J during manoeuvres

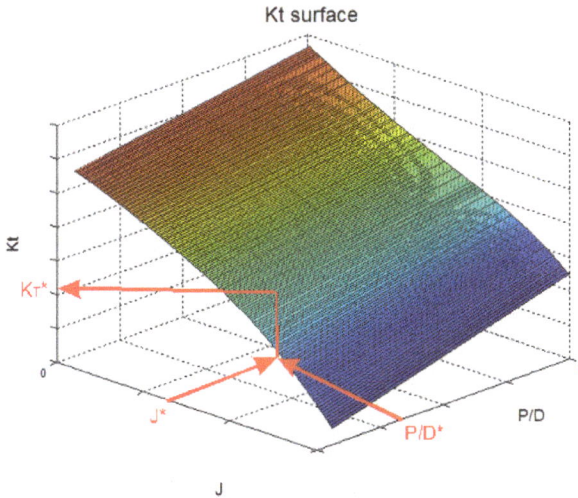

Figure 6.9: Kt surface

To evaluate the forces produced by the propeller along $\underline{b}_2$ and $\underline{b}_3$, the Ribner theory [36], based on semi-empirical formulations was used. This formulation relates the propeller hydrodynamic forces (non-dimensional) with their main geometrical characteristics.

The force is evaluated with the following formula:

$$F_{HD,b2} = \frac{k_s f(a)\sigma I_1}{1 + k_a \sigma I_1} \frac{A_0}{L^2} v_{prop(2)}^b$$

Where:

$f(a)$ is the q -factor:

$$f(a) = \frac{(1 + a)\left[(1 + a) + (1 + 2a)^2\right]}{1 + (1 + 2a)^2}$$

a is the inflow factor defined as follows:

$$a = \frac{1/2u_{A_\infty}}{u(1 - w)}$$

k_s is the spinner factor
k_a is the side wash factor
u_{A_∞} and u are defined in Paragraph 7.2
σ is the blade 'solidity'

$$\sigma = \frac{3Z}{4\pi}\left(\frac{ch_{0.75}}{D}\right)$$

I_1 is the side-area index:

$$I_1 = \frac{3}{4}m_0\frac{A_E}{Z}$$

m_0 is slope of the foil lift curve:

$$m_0 = 0.952\pi$$

By the same procedure, it is possible to evaluate the hydrodynamic force along $\underline{b}_3$:

$$F_{HD,b3} = \frac{k_sf(a)\sigma I_1}{1 + k_a\sigma I_1}\frac{A_0}{L^2}v_{prop(3)}^b$$

We assume that all blades develop the same thrust and require the same torque (the wake fraction field as a function of blade angular position is not considered):

$$\underline{F}_{HD\_blade} = \frac{\underline{F}_{HD}}{Z}$$

Where:

$\underline{F}_{HD\_blade}$ is the single blade developed force vector

The resultant hydrodynamic force is assumed to be acting on the hydrodynamic centre of the blade CH; the components of the vector in the **b**-basis are denoted by (x_{CH}, y_{CH}, z_{CH}). So, it is possible to evaluate the moment with the classical formula:

$$\underline{Q}_{SH} = (CH - O) \wedge \underline{F}_{HD}$$

With:

$$\tau_P = [\underline{F}_{HD}, \underline{Q}_{SH}]$$

The above resultant moment $\underline{Q}_{SH}$ needs to be decomposed according to the **f**-basis, then the component along the unit vector $\underline{f}_3$ can be evaluated to be used in the piston motion equation.

6.4.3 Frictional Forces

The blade seat on the hub is made by a bearing that supports forces in both axial and radial directions (Fig. 6.10). These frictional forces have been derived for each direction separately.

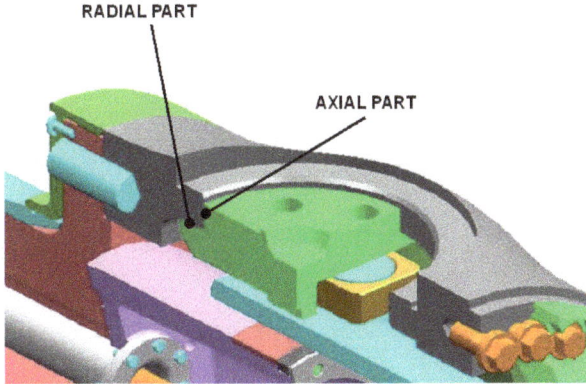

Figure 6.10: Blade seat

The frictional radial force and moment have been evaluated by the procedure proposed by Godjevac and co-authors [29], so it is not further discussed here.

The frictional axial force F_{AX} is the vectorial sum of the two components in the blade fixed frame **f** as shown in Fig. 6.11. The third component has been neglected for simplicity.

$$F_{AX} = \sqrt{F^2_{AX,f_1} + F^2_{AX,f_2}}$$

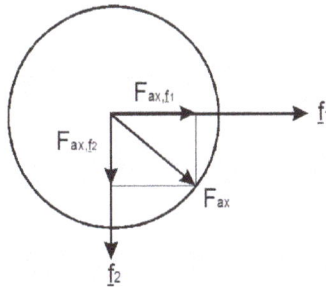

Figure 6.11: Axial contact

The interaction between the blade carrier and the hub is considered a point-wise contact. By using the static equilibrium of forces, the axial components along f_1 and f_2 can be assessed:

$$\sum F_{AX,f_1} = 0 \Rightarrow F_{HD,f_1} + F_{I,f_1} + F_{AX,f_1} = 0 \Rightarrow F_{AX,f_1} = -F_{HD,f_1} - F_{I,f_1}$$

$$\sum F_{AX,f_2} = 0 \Rightarrow F_{HD,f_2} + F_{I,f_2} + F_{AX,f_2} = 0 \Rightarrow F_{AX,f_2} = -F_{HD,f_2} - F_{I,f_2}$$

Where:

F_{HD,f_1} and F_{HD,f_2} are the hydrodynamic force components along f_1 and f_2, respectively

F_{I,f_1} and F_{I,f_2} are the inertial force components along f_1 and f_2, respectively

Then, the torque due to the axial force is given by:

$$Q_{AX} = \kappa_f F_{AX} d_a$$

Where:

d_a [m] is the distance between the point where the axial force is supposed to be applied and the hub friction seat of the propeller blade root

κ_f is the non-dimensional frictional coefficient of the blade and hub materials

The sum of the axial and radial friction torques along f_3 provides the total friction spindle torque Q_{SF}.

It possible to obtain τ_{PF} by the following relation:

$$\tau_{PF} = [F_{ax\_f1} + F_{rad\_f1}, \; F_{ax\_f2} + F_{rad\_f2} Q_{ax\_f3} + Q_{rad\_f3}]$$

6.4.4 Hydraulic Forces

The oil pressure needed to turn the blade or, alternatively, to hold it in the desired position is supplied by a hydraulic power pack consisting of pumps and valves. The actuating system consists of a double effect piston (yoke) with single rod and circular section. The developed force is proportional to the yoke area and to the oil pressure; a difference exists between the developed force in the forward and backward directions with the following relationship:

$$F_{hyd\_f} = A_1 p_1$$

$$F_{hyd\_b} = A_2 p_2$$

To evaluate the dynamic change of pressure, the following relationships are used:
− state equation:

$$\rho_{oil} = \rho_{i,oil} + \frac{\rho_{i,oil}}{B} p$$

− continuity equation:

$$\sum \dot{m}_i - \sum \dot{m}_o = \frac{d\left(\rho_{oil}V_c\right)}{dt} = \frac{d\rho_{oil}}{dt}V_c + \rho_{oil}\frac{dV_c}{dt}$$

Where:

V_c is the piston chamber volume

ρ_{oil} is the oil mass density

B is the oil Bulk modulus

$\dot{m}_i$ is the mass flow in

$\dot{m}_o$ is the mass flow out

By neglecting the dependence of the density on the temperature and combining the state equation with the continuity equation, we obtain:

$$q_i - q_o = \frac{V_c}{B}\frac{dp}{dt} + \frac{dV_c}{dt}$$

Where:

q_i is the volumetric flow in

q_o is the volumetric flow out

The leakage of the hydraulic actuator between the two chambers is evaluated by the following coefficient:

$$C_{ip} = \pi\frac{D_P}{2}\frac{t_0^3}{L_P}\frac{1}{6\mu}$$

Where:

D_P is the piston head diameter

t_0 is the orifice thickness

L_P is the thickness of the piston head

μ is the oil dynamic viscosity

Finally, the pressure differential equation becomes:

$$\dot{p}_i = \left(q_i - C_{ip}p_i - A_i\dot{x}_{pist}\right)\frac{B}{A_i x_{pist} + V_0}$$

Where:

V_0 is the initial chamber volume

7 Rudder

The rudder is the onboard machinery most widely used to manoeuvre the ship. It is a passive manoeuvring device that uses flow generated by the surge ship motion to develop a suitable lateral force and an associated moment, which are responsible for the ship turns.

The rudder force can be decomposed into two different components:

- The first is perpendicular to the fluid velocity direction, and it is called Lift.
- The second is parallel to the fluid velocity direction, and it is called Drag.

It is possible to assume that both lift and drag forces are applied in the pressure centre of the rudder.

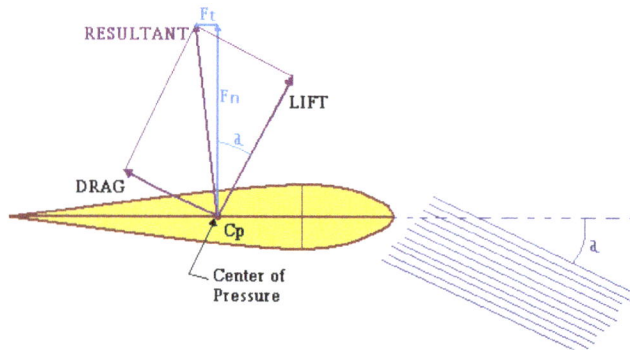

Figure 7.1: Rudder Force

The commonly used rudders have a decided elongation in one of their two dimensions and generally have a rounded nose contour at the leading edge and a pointed trailing edge. The main features that characterize a rudder are:

- Contour Type: i.e. NACA, HSVA, IFS, etc.
- Mean line: the line that links the trailing edge and the leading edge
- Attack angle (δ_{eff}): the angle between the mean line and the flow velocity direction
- Average chord (c): the mean of the curves length, linking the trailing edge and the leading edge (chords), evaluated at root and at tip
- Average span (s): the mean rudder height
- Maximum thickness (t): the maximum rudder thickness
- Rudder Area (A_R): the area invested by the fluid
- Aspect Ratio ($\frac{s}{c}$): the ratio between the span and the chord

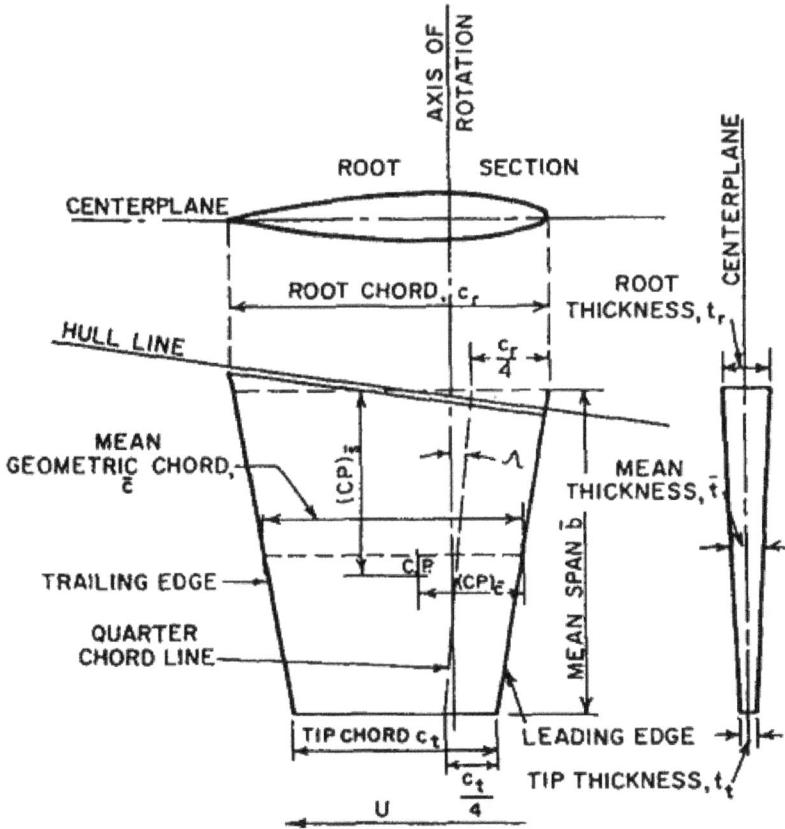

Figure 7.2: Rudder overview

Usually, the rudder is installed at the ship stern in to maximise the force arm with respect to the centre of gravity; furthermore, in this position the rudder can also exploit the flow acceleration generated by the propeller. For example, if the ship speed is zero or close to zero and the propeller works in bollard condition, the rudder can generate a hydrodynamic force (if located in the propeller slipstream).

After the above brief description of the main features of a rudder, the main aim is to develop a three dimensional model to describe the main effects that the rudder produces during a ship manoeuvre. Another important goal is the development of a model independent of the specific characteristics of the particular rudder one can consider. To do this, all involved variables have must be used in non-dimensional form according to Table 7.1 and Table 5.1.

Moreover, to implement the rudder model, it has been necessary to deal with the flow field in the stern region of a ship taking into account the mutual interference among rudder, hull, and propeller; in addition, the rudder has been modeled taking into account all six ship motions.

Table 7.1: Scaling factor

VARIABLES	UNITS (I.S.)	SCALING FACTOR
Rudder position & main geometric characteristic	[m]	L
Lift & Drag Forces	[N]	$\frac{1}{2}\rho L^2 V^2$
Lift & Drag Moments	[Nm]	$\frac{1}{2}\rho L^3 V^2$
Linear velocity	[m/s]	V
Angular Velocity	[rad/s]	$\frac{V}{L}$

Usually, the rudder's main function is to correct the heading of a ship; however, depending on the ship type, the rudder can also be used to produce or correct roll motion. In recent years the rudder roll system has been developed, and nowadays applications are being implemented, especially in the naval field. The complete knowledge of the rudder forces and moments allows better system design, and in particular it allows the use of simulation techniques to develop and tune the rudder roll control system before its installation onboard.

In the following sections, the main characteristics of the proposed rudder mathematical model are described with the aim to provide a theoretical knowledge about the principal hydrodynamic phenomena involved. This model is valid for forward ship velocity. The rudder taken into account for this formulation is a Spade type.

7.1 Reference Frame

In order to obtain a correct model development, it has been necessary to introduce all the coordinate systems involved in the formulation of our model:
– Rudder at zero degrees fixed frame (**d**)

The **d**-frame $\left(O_d, \underline{d}_1, \underline{d}_2, \underline{d}_3\right)$, shown in Fig. 7.3, is fixed to the rudder (at 0°rudder angle). The unit vector $\underline{d}_1$ points towards the leading edge along the chord when the rudder is in its 0°position; $\underline{d}_3$ has direction parallel to the rudder surface and points downwards; $\underline{d}_2$ is the result of the vector product $\underline{d}_2 = \underline{d}_3 \wedge \underline{d}_1$. The origin O_d of the frame is positioned at a fixed point in the rudder (that is considered as fixed for the sake of simplicity) more precisely it is assumed that O_d location is given by:

Where $x_{rud}, y_{rud}, z_{rud}$ are the coordinates of the rudder root with the respect to the pole O_b. δ_{fix} denotes the rudder axis inclination; in this formulation, it is arbitrarily assumed to locate the centre of the system at 45% of the span, a vertical position near the effective centre of pressure. In case a better approximation is necessary, further shifts of the force may be applied. The rudder is installed onboard with a transversal inclination to take advantage in terms of roll moment deduction.

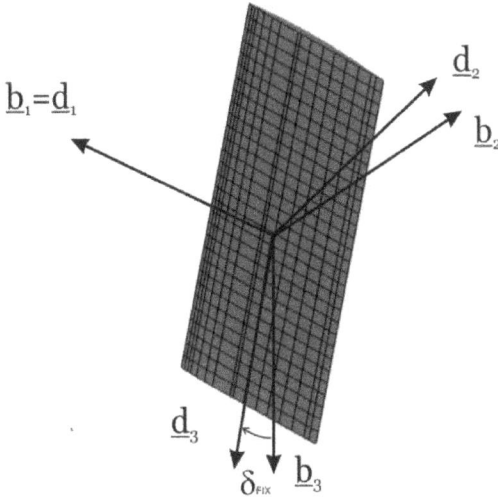

Figure 7.3: Rudder at zero degrees Fixed Frame (**d**)

The linear transformation between the **b**-basis and the **d**-basis can be expressed as:

$$\underline{d}_i = \sum_{j=1}^{3} \left(R_b^d\right)_{ij} \underline{b}_j$$

where the orthogonal matrix R_b^d is function of the fixed angle δ_{fix}:

$$R_b^d = \begin{bmatrix} 1 & 0 & 0 \\ 0 & \cos(\delta_{fix}) & -\sin(\delta_{fix}) \\ 0 & \sin(\delta_{fix}) & \cos(\delta_{fix}) \end{bmatrix}$$

– Rudder-fixed frame (**w**)

The **w**-frame $\left(O_w, \underline{w}_1, \underline{w}_2, \underline{w}_3\right)$, shown in Fig. 7.4, is fixed to the rudder (moving altogether with it). The positive unit vector $\underline{w}_1$ points towards the leading edge and it is rotated by δ with the respect to $\underline{d}_1$. δ is the instantaneous rudder angle. $\underline{w}_3$ is equal to $\underline{d}_3$, and $\underline{w}_2$ is the result of the vector product $\underline{w}_2 = \underline{w}_3 \wedge \underline{w}_1$. The origin O_w of the **w**-frame is assumed to coincide with O_d. Theoretically, a moving pole O_w should be employed to consider the shift of the centre of pressure of the rudder as a function of the rudder angle; however, choosing $O_w = O_d$ involves only a small and negligible error in terms of global manoeuvring performances. This new reference frame allows the forces and fluid velocity components to be expressed as functions of the rudder angle δ (a positive rudder angle corresponds to a portside ship turn).

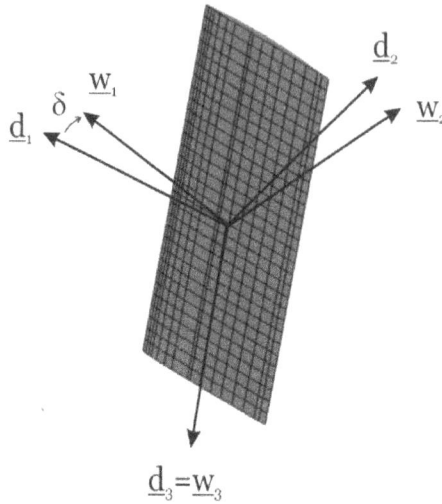

Figure 7.4: Rudder-Fixed Frame (**w**)

The linear transformation between the **d**-basis and the **w**-basis can be expressed as:

$$\underline{w}_i = \sum_{j=1}^{3} \left(R_d^w\right)_{ij} \underline{d}_j$$

where the orthogonal matrix R_d^w is function of the variable rudder angle δ:

$$R_d^w = \begin{bmatrix} cos(\delta) & sin(\delta) & 0 \\ -sin(\delta) & cos(\delta) & 0 \\ 0 & 0 & 1 \end{bmatrix}$$

– Fluid-fixed frame (**z**)

The **z**-frame, shown in Fig. 7.5, $\left(O_z, \underline{z}_1, \underline{z}_2, \underline{z}_3\right)$ is fixed to the fluid velocity vector, incoming on the rudder; this is obviously only an ideal direction, which is used to evaluate the effective rudder angle and consequently the rudder forces as obviously the flow direction at rudder location during manoeuvres is not unique. The unit vector $\underline{z}_1$ points towards the flow velocity direction at rudder and it is rotated by δ_{EFF} with respect to $\underline{w}_1$. $\underline{z}_3$ is equal to $\underline{w}_3$, and $\underline{z}_2$ is the vector product $\underline{z}_2 = \underline{z}_3 \wedge \underline{z}_1$. The origin O_z of the frame coincides with O_w. This reference frame allows the Lift and the Drag forces to be evaluated taking into account the effective angle of attack δ_{EFF} between the fluid and the rudder.

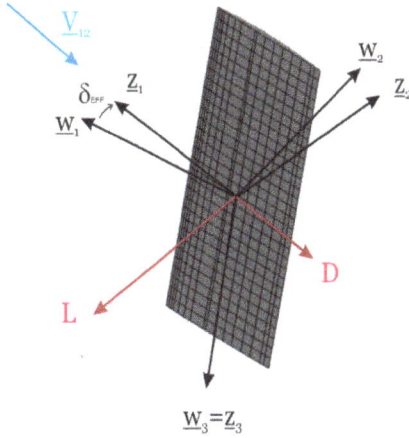

Figure 7.5: Fluid-Fixed Frame (**z**)

The linear transformation between the **z**-basis and the **w**-basis can be expressed as

$$\underline{z}_i = \sum_{i=1}^{3} \left(R_w^z\right)_{ij} \underline{w}_j$$

Where the orthogonal matrix R_w^z is function of the variable angle δ_{EFF}:

$$R_w^z = \begin{bmatrix} \cos(\delta_{eff}) & \sin(\delta_{eff}) & 0 \\ -\sin(\delta_{eff}) & \cos(\delta_{eff}) & 0 \\ 0 & 0 & 1 \end{bmatrix}$$

7.2 Hull – propeller – rudder interactions

The hull presence alters the fluid velocity vector in the stern zone where the rudders are located. The disturbances due to the hull on the incoming flow (the hull has a straightening effect on the flow) have been evaluated with statistical regressions derived from experimental tests. To account for these effects, the straightening effect has been decomposed in two components [11]: one component acts on the sway motion and the other acts on the yaw motion. Therefore, two correction factors have been introduced to modify the fluid velocity vector.

In addition, a third correction factor has been introduced, which modifies the roll motion because in our case we are considering six degrees of freedom.

The relation between the components of linear and angular ship velocities evaluated in 0_b and the components of the same velocities evaluated in the stern region before the propeller is assumed to be:

$$v_{ster} = vy$$

With:

$$y = \begin{bmatrix} 1 & y_v & 1 & y_p & 1 & y_r \end{bmatrix}$$

y is called fluid straightening vector.

In more detail, according to Ankudinov regression, the coefficients of y can be evaluated through the following formula:

$$y_v = 1 - 1.5 \cdot (B_v/L) \cdot \left(C_B \cdot B_v/T_s\right)^{0.5}$$

$$y_r = y_v \cdot (0.44 + 0.22 \cdot y_v)$$

$$y_p = y_v$$

Where:

B_v is the vessel breadth

C_B is the block coefficient

T_S is the ship draft

The fluid velocity evaluated in O_d is different from the fluid velocity evaluated in O_b not only because of the flow straightening, but also because of the different position. So, the rigid motion formula has been used to calculate the local (rudder) fluid velocity:

$$\underline{v}^b_{rud}\,(1:3) = \underline{v}^b_{rud}\,(1:3) + \underline{v}^b_{rud}(4:6) \wedge (O_b - O_d)$$

There is also a strong interaction between the propeller and the rudder; the longitudinal velocity component should be evaluated taking into account the longitudinal velocity at the rudder position due to the ship motions (ship longitudinal velocity modified by the asymmetrical wake fraction)in addition to the propeller accelerated flow considering rudder position relative to propeller axis, slipstream diameter reduction, and flow turbulence corrections.

These effects cannot be neglected without compromising the accuracy of the rudder force evaluation.

The acceleration effect is evaluated by the actuator disk theory; the velocity at an infinite distance downstream from the rudder is:

$$u_{A\infty} = \sqrt{(1-w)^2\, u^2 + \frac{8}{\pi} K_T \,(nD)^2} - (1-w)u$$

Then the velocity in an upstream region is added:

$$u_\infty = u_{A\infty} + (1-w)u$$

The velocity evaluated in the propeller disk plane is:

$$u_{prop} = \frac{1}{2}u_{A\infty} + (1-w)u$$

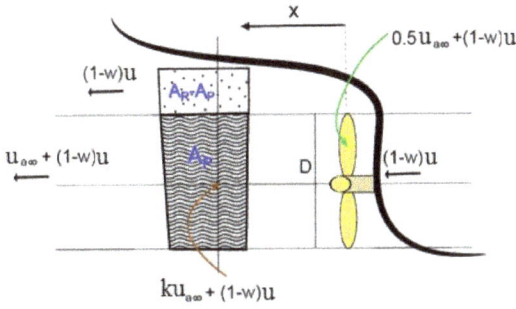

Figure 7.6: Rudder-Propeller Interaction

So, it is possible to evaluate the surge flow velocity at every point of interest using the following equation derived from actuator disk theory:

$$u(x) = k_{prop}(x)u_{A\infty} + (1 - w)u$$

Where k_{prop} is a coefficient that provides the accelerated velocity as a function of the non-dimensional distance between the rudder and the propeller; this is obtained using the diagram in Fig. 7.7.

$$k_{prop} = 0.5 + 0.36\tanh\left(0.98d_{prop\_rud}\right) + 0.11\tanh(15d_{prop\_rud})$$

where d_{prop_rud} is the non-dimensional distance (with respect to $D/2$), along $\underline{b}_1$ direction, between the rudder and the propeller.

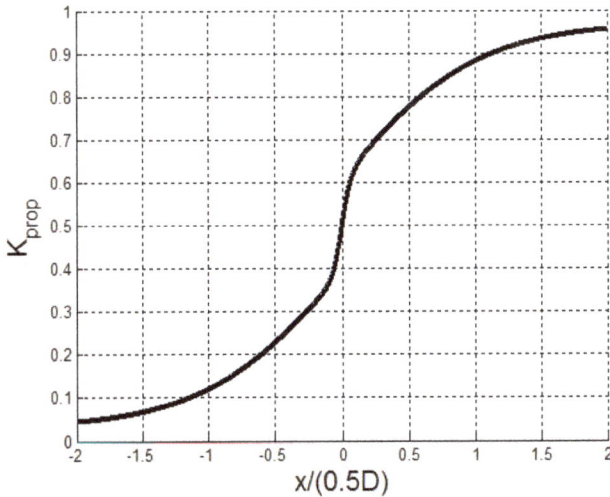

Figure 7.7: k_{prop} coefficient

The last correction for the surge velocity was introduced to take into account the experimental results about the rudder/propeller complex. In particular, on the basis of Molland's work [31], a new coefficient k_{slip} was introduced that reduces the effect due to the propeller slipstream. In this way, the first component of the velocity vector becomes:

$$u(x) = k_{prop}(x) \cdot u_{A\infty} \cdot k_{slip} + (1 - w) \cdot u$$

Once the velocity is known, it is possible to define the flow tube diameter at a certain location x using the flow conservation equation:

$$D_T(x) = D\sqrt{\frac{u_{prop}}{u(x)}}$$

However, the hypothesis of flow conservation is not satisfactory because there is a mixing between the flow inside the propeller slipstream and the surrounding flow. To account for this effect, the following formula was applied [18]:

$$\Delta r(x) = 0.15x\frac{u(x) - u(1 - w)}{u(x) + u(1 - w)}$$

Where:

Δr is the radius increment

In this way, the surge velocity is corrected as follows:

$$u_{corr}(x) = [u(x) - u(1 - w)]\left(\frac{r}{r + \Delta r}\right)^2 + u(1 - w)$$

Now, it is possible to calculate the total velocity through weighted average between the accelerated flow velocity on the rudder inside the flow tube and the undisturbed flow velocity outside the flow tube:

$$u_{acc}^b = \frac{A_P}{A_R} \cdot [u_{corr}(x)]^2 + \frac{A_R - A_P}{A_R}[u(1 - w)]^2$$

Where:

A_P is the rudder area inside the flux tube.

7.3 Proposed Forces Evaluation

Using a composition of rotational matrices, it is possible to represent the rudder velocity vector in the **w**-basis:

$$\underline{v}_{rud}^w(1:3) = R_w^d \times R_d^b \times \underline{v}_{rud}^b(1:3)$$

To evaluate the forces developed by the rudder, it is necessary to consider the real angle between the rudder and the fluid; this angle is obtainable by the following equation:

$$\delta_{eff} = \text{atan}\left(\frac{v_{rud}^w(2)}{u_{acc}^w}\right)$$

We also remember an alternative definition of the attack angle:

$$\delta_{eff} = \delta - \beta_{loc}$$

Where:

β_{loc} is the local drift angle

$$\beta_{loc} = \mathrm{atan}\left(\frac{v_{rud}^d(2)}{u_{rud}^d}\right)$$

$v_{rud}^f(2)$ is the sway velocity evaluated in the rudder position with respect to the **d**-frame, taking into account the fluid straightening due to the hull

u_{rud}^d is the accelerated surge velocity evaluated in the **d**-frame

It is also possible to calculate the velocity module in the plane$< \underline{z}_1\underline{z}_2 >$. Using the hypothesis that the velocity component along $\underline{z}_3$ is negligible, we obtain:

$$V_{z1z2} = \sqrt{v_{rud}^w(1)^2 + v_{rud}^w(2)^2}$$

Subsequently, the forces acting on the rudder, expressed in **z**-basis, can be evaluated. Rudder forces are evaluated considering rudder open water characteristics, properly modified by hydrodynamic interaction with propeller and hull and the rudder type. The lift slope coefficient is expressed through the following formula [26]:

$$dC_L = \frac{1.95 \cdot \pi}{1 + \frac{3}{A.R._{eff}}}$$

Where:

AR_{eff} is the effective aspect ratio

AR_{geom} is the geometric aspect ratio, defined as follows:

$$AR_{eff} = 2AR_{geom} = 2\frac{S}{C}$$

The effective aspect ratio was introduced because the rudder root is quite close to the hull; in this configuration, cross-flow effects on one extremity of the rudder are blocked, so the rudder aspect ratio may be doubled with respect to a rudder with two free extremities. Cross-flow is a 3-D effect generated by significant differences in pressure on the two side of the airfoil. This pressure field allows fluid to cross from the high pressure zone to the low pressure zone with a consequent drop in performance.

To assess the drag force in non-dimensional form, the following formula [45] (with a correction) was adopted:

$$D_{rud} = -\mathrm{sign}(v_{rud}^b(1))\left[\frac{(dC_L\,\delta_{EFF})^2}{e_{Osw}\,\pi\,AR_{eff}} + C_{D0}\right] AR\frac{V_{z1z2}^2}{L^2V^2}\,corr_{CD\_spade}$$

Where:

e_{Osw} is the Oswald correction shape factor that takes into account the non-elliptical circulation around the rudder

C_{D0} is the drag coefficient for $\delta_{EFF} = 0$

$corr_{CD\_spade}$ is an empirical corrective coefficient obtained by the following equation as a function of the propeller hydrodynamic load:

$$corr_{CD\_spade} = \left(-0.29181 \left(\frac{K_T}{J^2} \right)^2 + 0.82005 \left(\frac{K_T}{J^2} \right) + 1.23509 \right)$$

The lift coefficient C_L is defined in three different ranges of rudder angle by proper laws; these ranges depend on the stall angle and the peak angle.

The stall angle α_{STALL} is the effective rudder angle where the stall phenomena starts and is obtained by linear interpolation of the function of the propeller hydrodynamic load K_T/J^2 using the follow diagram:

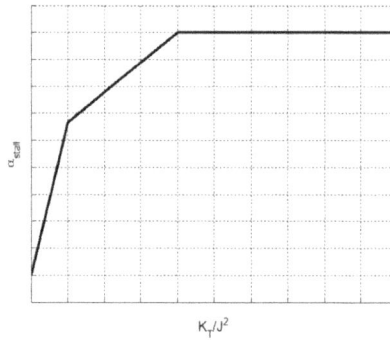

Figure 7.8: Diagram of α_{STALL}

The angle α_{PEAK} is the effective rudder angle at which the maximum lift is produced and is obtained by linear interpolation of a function of the α_{STALL} angle using the following diagram:

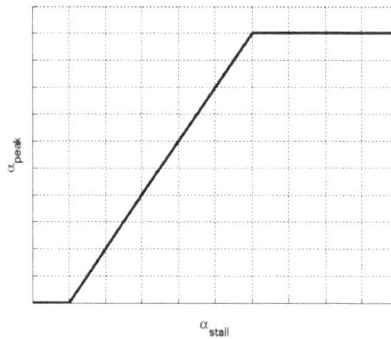

Figure 7.9: Diagram of α_{PEAK}

The corresponding lift curve is outlined in the following figure:

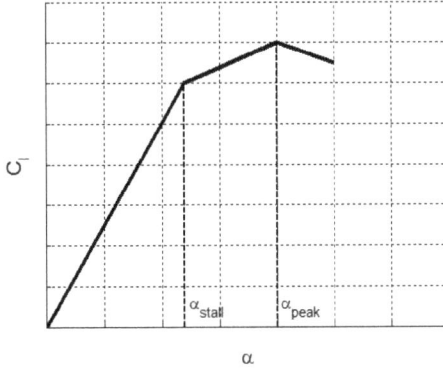

Figure 7.10: Trend of lift coefficient curve

In more detail, the Lift is expressed in non-dimensional form through a piecewise function:

If $\delta_{EFF} < \alpha_{STALL}$

$$L_{rud} = dC_L \, |\delta_{EFF}| \frac{A_R \, V^2_{z1z2}}{L^2 \, V^2}$$

If $\alpha_{STALL} \leq \delta_{EFF} \leq \alpha_{PEAK}$

$$L_{rud} = \left[dC_L \alpha_{STALL} + dC_L \left(|\delta_{EFF}| - \alpha_{STALL}\right) \frac{C_{L\_STALL}}{range\_stall} \right] \frac{A_R \, V^2_{z1z2}}{L^2 V^2}$$

If $\delta_{EFF} > \alpha_{PEAK}$

$$L_{rud} = \left[dC_L \alpha_{STALL} + dC_L \left(\alpha_{PEAK} - \alpha_{STALL}\right) \frac{C_{L\_STALL\_SPADA}}{range\_stall} \right.$$
$$\left. - dC_L \left(|\delta_{EFF}| - \alpha_{PEAK}\right) \frac{C_{L\_STALL\_SPADA}}{range\_stall} \right] \frac{A_R \, V^2_{z1z2}}{L^2 V^2}$$

Where:

C_{L_STALL} is a semi-empirical value

$range\_stall$ is a semi-empirical value

So, it is possible to compose the forces vector developed by the rudder:

$$\underline{F}^z_{rud} = (D_{rud} L_{rud}, 0)$$

Now, it is possible to express the forces developed by the rudder in the **b**-basis using the (transposes of the) rotational matrices introduced above:

$$\underline{\tau}_r(1:3) = \left(R^b_d \cdot R^d_w \cdot R^w_z \right) \underline{F}^z_{rud}$$

It is also possible to calculate the moments generated by the rudder introducing the rudder position, identified by the distance between the pole O_b and the pole O_d, and applying the following formula:

$$\underline{\tau}_r(4:6) = (O_b - O_d) \wedge \underline{\tau}_r(1:3)$$

Finally, after this procedure, we can add the rudder forces and moments in the ship non-dimensional motion equations.

8 Validation

In order to validate the proposed methodologies and the software platform, a ship equipped with CPPs driven by a gas turbine has been modeled, and an experimental campaign has been performed.

For confidentiality reasons, the obtained ship data cannot be published, and all the results are presented in non-dimensional form. The main ship characteristics are reported in the following table:

Table 8.1: Main ship characteristics

L/B	7
B/T	4
F_N	0.37
Propeller	5 bladed CPP
Rudder	Rudder roll
Main Engine	Gas Turbine

8.1 Comparison

During the sea trials, the main propulsion variables and ship manoeuvring characteristics have been recorded by the on-board automation system and by a supplementary acquisition system, which was suitably installed on-board to record the typical manoeuvrability parameters. In total, more than 60 channels have been recorded and post-processed. Hereinafter a comparison between experimental measurements and simulated data for different manoeuvres is reported, but for the sake of brevity, only the most relevant results are shown. Both kinds of data have been sampled every single second. To verify a good correlation between sea trials and simulation results, a slam start, a tight turning circle, and the zig-zag manoeuvre have been analysed.

8.1.1 Slam Start

Reported below is a comparison between experimental and simulated data for a slam start manoeuvre. Slam start is an emergency manoeuvre used to determine the propulsion plant performance and its control system reliability during tough manoeuvres. It is also a necessary trial to test and validate the new control logic introduced in 4.2.1. Starting from zero speed (or around zero speed), the ship is brought to the maximum speed reachable by the current propulsive configuration.

All experimental data are sampled every second. The simulated output data are sampled every second. The propulsion configuration during the sea trial was in electric mode. In the simulation model, the final propulsion control system configuration installed on board is not present.

Fig. 8.1 shows the command lever position time history for both port and starboard step levers. Each value is dimensionless with respect to the maximum lever position. This figure demonstrates that the simulated manoeuvre corresponds to the experimental one.

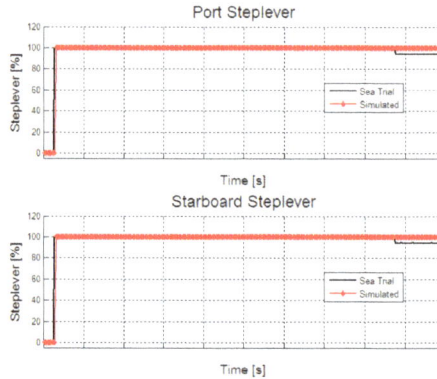

Figure 8.1: Slam Start – Step Lever vs. Time

Fig. 8.2 shows the ship speed time history. Each value is dimensionless with respect to the maximum speed reachable in electric mode configuration. In the experimental curve, a step behaviour is observed. This is because the variable has been recorded without decimal values. A very good correlation between the two curves can be seen. This demonstrates that the simulation platform is able to represent the real ship behaviour during a high-dynamic transient. This result means that all macromodels (propulsion plant, control system, and manoeuvrability) and their mutual interactions are correctly schematized.

Fig. 8.3 shows the revolution regime time history for both transmission lines, port and starboard. Each value is dimensionless with respect to the maximum rpm commanded in electric mode configuration. The two curves are completely overlapped, and only a few differences occur in the first part of the manoeuvre. This could be due to a small difference of the rpm controller proportional gain. The system appears to be completely controlled. No rpm overshoots appear, and the manoeuvre has been performed without any interference of the electronic control protection.

Fig. 8.4 shows the propeller pitch angle over time for both shaft lines, port and starboard. Each value is dimensionless with respect to the design propeller pitch angle. Some differences during transients occur. The simulated pitch change is quicker

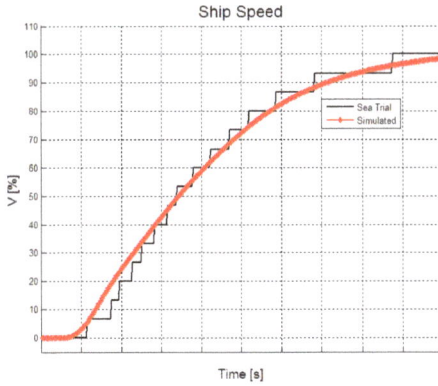

Figure 8.2: Slam Start – Ship Speed vs. Time

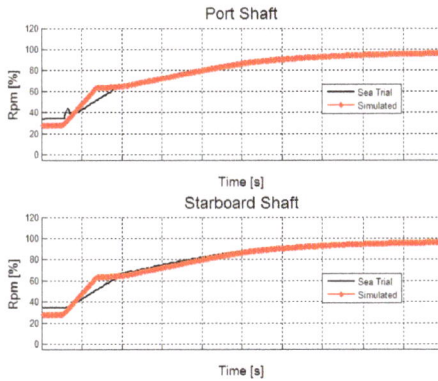

Figure 8.3: Slam Start – Shaft lines revolution vs. Time

than the real one. Some real system delays (electric transmission line delay, PLCs inner time cycle, etc.) can create this difference or maybe a different tuning from the real PLCs installed onboard exists.

Fig. 8.5 shows the oil pressure inside the piston chamber, for port and starboard pitch change mechanisms, over time. Each value (both simulated and experimental) is dimensionless with respect to the maximum portside oil pressure reached during the trial.

The two curves show a different initial slope. The simulated oil pressure rises more slowly than the experimental one. This could be caused by a different positioning of the measuring equipment. In fact, the simulated oil pressure is evaluated in the actuating chamber and remains constant for the entirety of the manoeuvre. The experimental data is probably measured after the pump. When the blade stops moving, a pressure drop occurs because the valve that isolates the actuating chamber has been closed.

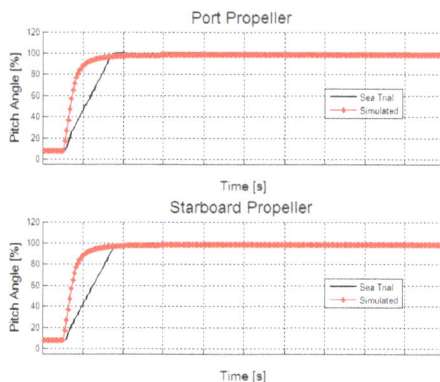

Figure 8.4: Slam Start – Propellers pitch angle vs. Time

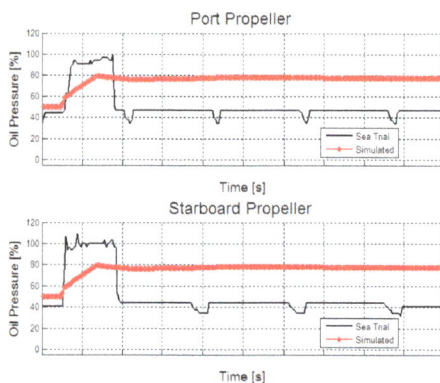

Figure 8.5: Slam Start – Oil pressure vs. Time

The maximum value calculated through the simulator is 20% less than the experimental maximum value.

Fig. 8.6 shows the propeller torque over time for both shaft lines, port and starboard. Each value is dimensionless with respect to the steady state value of the end of the manoeuvre.

The simulated data rises more quickly than the experimental data. This is an effect of the propeller pitch change velocity shown in Fig. 8.4. It is interesting to note that, during the transient, 110% of the torque at the end of the manoeuvre is reached.

Fig. 8.7 shows the propeller thrust over time for both shaft lines, port and starboard.

Each value is dimensionless with respect to the steady state value of the end of the manoeuvre.

The thrust measurement is difficult to measure, and the following curves have high values for validating the propeller open water characteristics and the methodology described in 6.4.2. The simulated data rises more quickly than the experimental data. This is an effect of the propeller pitch change velocity, which is shown in Fig. 8.4. It is interesting to note that, during the transient, there occurs 140% of the thrust at the end of the manoeuvre. This effect in Gas Turbine mode can be dangerous for the thrust bearing. There is an optimal correlation between the curves.

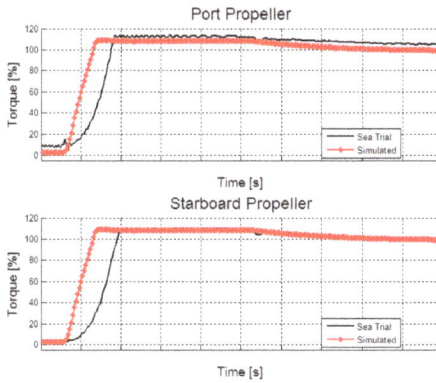

Figure 8.6: Slam Start – Shafts Torque vs. Time

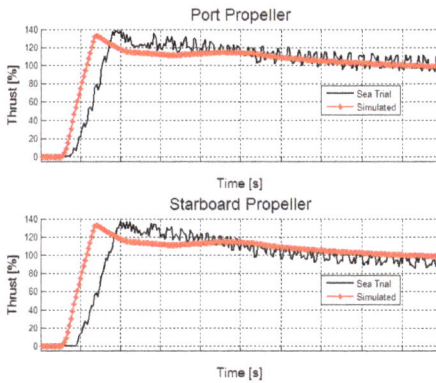

Figure 8.7: Slam Start – Propellers Thrust vs. Time

Fig. 8.8 shows the electric propulsion motor power over time for both shaft lines, port and starboard. Each value is dimensionless with respect to the maximum power that the electric motor can generate. There is a good overlap between the two curves. Small differences exist: the simulated motor dynamic is a little bit quicker than the real one in the initial period, and then the steepness becomes comparable just like

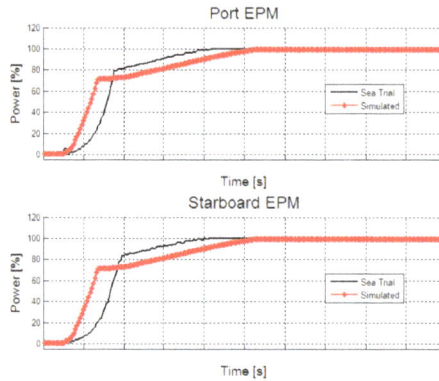

Figure 8.8: Slam Start – Electric Motors Power vs. Time

the shaft torque and the propeller thrust shown previously. This could be due to a fast propeller pitch change.

8.1.2 Turning

Below a comparison between experimental and simulated data for an evolution manoeuvre in a gas turbine configuration has been reported. All experimental data was sampled every ten seconds, whereas the simulated data was sampled every second. The considered turning circle manoeuvre has been performed at the maximum speed with a rudder angle near the maximum. In particular, the manoeuvre is characterized by an approach phase in straight motion, followed by a turning circle at constant rudder angle. A turning test is usually used to determine the effectiveness of the rudder, to assess steady-state turning characteristics, and to analyse the propulsion plant behaviour in transient conditions. During this manoeuvre, significant transversal heeling angles of the vessel may be reached, especially at high speeds. The propeller pitch angle has been recorded so that the effect of the control system during this manoeuvre can be completely validated. Furthermore, the propeller thrusts have been recorded, so it is been possible to also validate the methodology employed to represent asymmetric propellers loads.

Fig. 8.9 shows the commanded values of the rudder angle over time. Each rudder value is dimensionless with respect to the maximum rudder angle. This figure demonstrates that the simulated manoeuvre corresponds, in terms of rudder actuating velocity, to the experimental one.

Fig. 8.10 shows the ship trajectory in terms of non-dimensional position: X/L and Y/L. The simulation platform shows good precision with small differences on the tactical diameter below the 40% ship length. From this point of view, it is remarkable that the hydrodynamic coefficients, obtained by the adopted regression formula, have

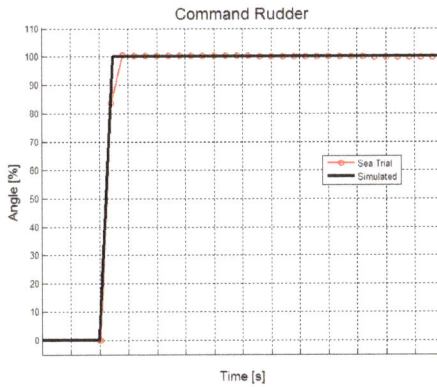

Figure 8.9: Turning - Rudder vs. Time

been further tuned on the basis of the model scale tests since particular attention is given to the effect of the ship's manoeuvre on the propulsion system.

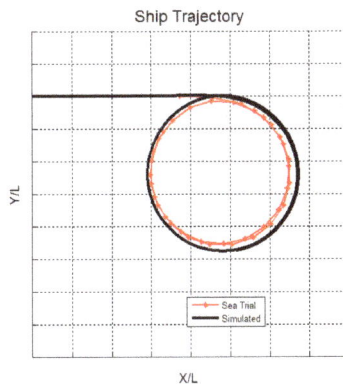

Figure 8.10: Turning – Trajectory

Fig. 8.11 shows the shaft power time history in both transmission lines, port and starboard. Each power value is dimensionless with respect to the initial manoeuvre power. Some interesting results appear from this figure. During the initial period of the manoeuvre, it is possible to see different behaviour in the two shaft lines. In the inner shaft line (port), there is a power reduction; in the outer shaft line (starboard), there is a power increase. During the stable period of the manoeuvre, both shaft lines require, with different magnitude, a higher power with respect to a straight course. These effects can be explained by the asymmetrical propeller flow described in 6.4.2.

Fig. 8.12 shows the revolution regime time history for both transmission lines, port and starboard. Each value is dimensionless with respect to the initial manoeuvre port

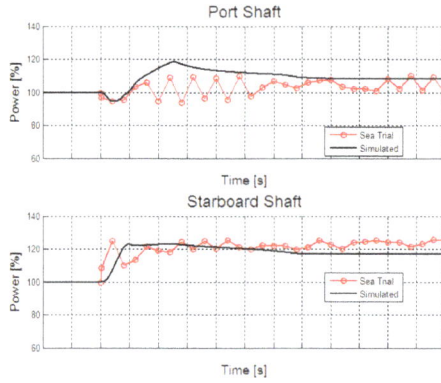

Figure 8.11: Turning – Shafts Power vs. Time

rpm. Some differences occur. During the sea trial, the rpm are more or less constant for the entirety of the manoeuvre. Instead, in the simulated time history, a symmetrical decrease of the shaft lines revolution of about 5% is observed. The symmetry is due to the particular gearbox installed onboard. Different reasons can additionally explain these discrepancies. In fact, during a manoeuvre, an increase in propeller torque is required. In the simulated manoeuvre, the gas turbine is not able to provide the necessary torque, giving rise to an rpm drop. The gas turbine model has been validated with manufacturer data, so it is possible that there are different thresholds for the over-torque control.

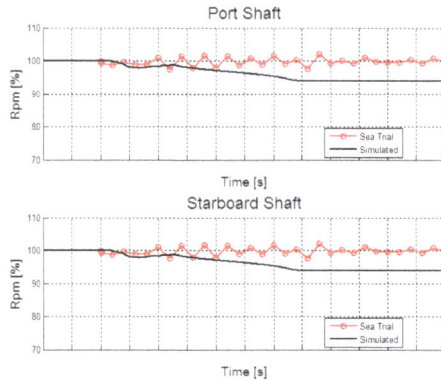

Figure 8.12: Turning – Shaft lines revolution vs. Time

Fig. 8.13 shows the ship speed time history. Each value is dimensionless with respect to the approach speed. When the rudder was turned portside, the ship speed decreased, reaching, when the manoeuvre stabilized, a value 35% lower than that in the beginning. A small difference of less than 5% between experimental records and simulation values is noted. Moreover, ship speed reduction is faster in the simulation. This behaviour might be partly ascribed to some differences in the automation settings and partly to residual differences in the manoeuvrability model.

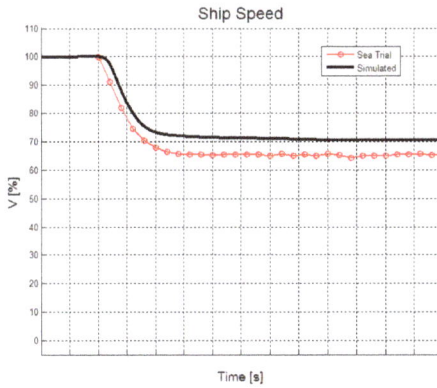

Figure 8.13: Turning – Ship speed vs. Time

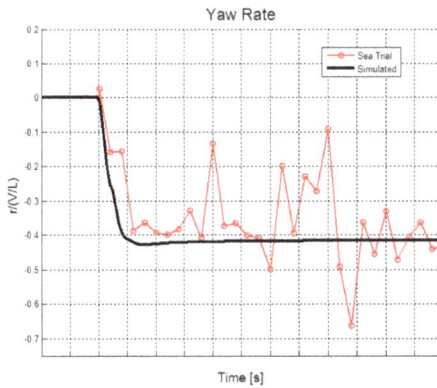

Figure 8.14: Turning – Yaw rate vs. Time

Fig. 8.14 shows the yaw rate time history. Each value is dimensionless with respect to the V/L ratio. The experimental data are scattered; however, the simulated are smoothened. The trends appear to be similar.

Fig. 8.15 shows the transversal heeling over time. Each numerical value is dimensionless with respect to the peak angle reached during this manoeuvre. A reasonably good matching between the two time histories is observed. This proves the validity of

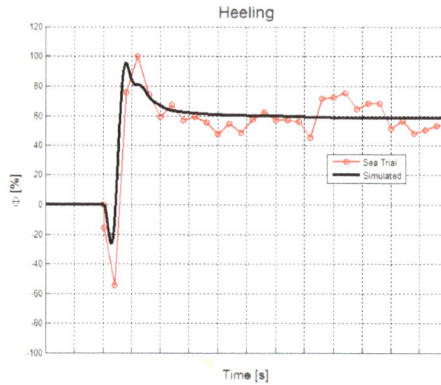

Figure 8.15: Turning – Heeling angle vs. Time

the evaluation of the forces acting on hull and rudder. In fact, a small difference of less than 10% on the first peak value and on the stabilized heel angle is observed; on the contrary, simulation is not able to capture the oscillations that are present in the sea trials records. The presence of these oscillations has yet to be investigated and might be ascribed to an external disturbance. This proves the validity of the manoeuvring model described in 5.3 and 5.4. Only a small difference in the peak values still exist.

8.1.3 Zig Zag

Next we report a comparison between experimental and simulated data for a Zig Zag manoeuvre. This type of manoeuvre is performed by setting the rudder from its initial central position to an assigned angle (usually 10 or 20 degrees) to starboard (or port). Once this angle has been reached, the rudder is kept steady until the ship has turned by the same angle to starboard (or port).

All experimental data were sampled every ten seconds, the simulated data were sampled every second. This manoeuvre has been performed at the maximum speed. The propulsion configuration was set in Gas Turbine mode. Zig Zag test is used to determine the ship course stability and to analyse the propulsion plant and its controller behaviour during transient periods. Unfortunately, the propeller pitch angle was not recorded during this sea trial.

Fig. 8.16 shows the command rudder angle versus time. Each rudder value is dimensionless with respect to the maximum rudder angle. This figure demonstrates that the simulated manoeuvre corresponds, in terms of rudder actuating velocity and timing, to the experimental one.

Fig. 8.17 shows the ship trajectory in terms of non-dimensional position: X/L and Y/L. A generallly good correlation, with very few differences below the 0.1 ship length, is observed.

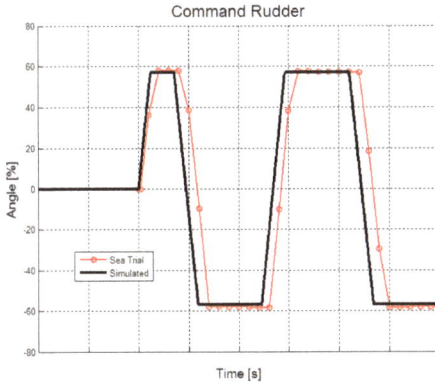

Figure 8.16: Zig Zag - Rudder vs. Time

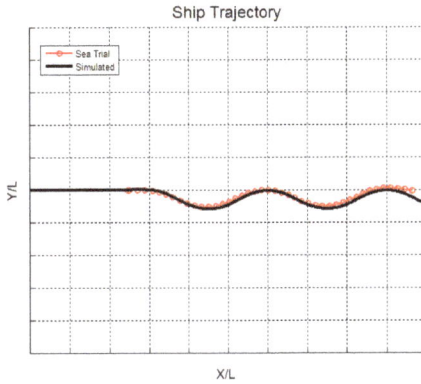

Figure 8.17: Zig Zag – Trajectory

Fig. 8.18 shows the shaft power time history in both transmission lines, port and starboard. Each power value is dimensionless with respect to the initial manoeuvre power. Some interesting results appear from this figure. During the entirety of the manoeuvre, different behaviour is observed on the two shaft lines. A sinusoidal behaviour appears; these effects can be charged to the asymmetrical propeller flow, described in 6.4.2, and to the propulsion control system effect on the propeller pitch angle reduction, described in 4.1.3. Unfortunately, it is impossible to prove the effective pitch reduction due to lack of recorded data. The experimental data sample time does not allow us to make any further conclusions regarding the different maximum power measurements for port shaft (105%) and starboard shaft (120%).

Fig. 8.19 shows the shaft line revolution regime time history for both transmission lines, port and starboard. Each value is dimensionless with respect to the initial manoeuvre rpm. During the sea trial, the rpm are more or less constant for the entire trial. Then, the variables start to oscillate; this phenomena might be due to a control

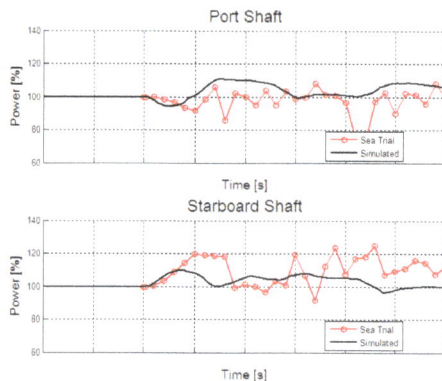

Figure 8.18: Zig Zag – Shafts Power vs. Time

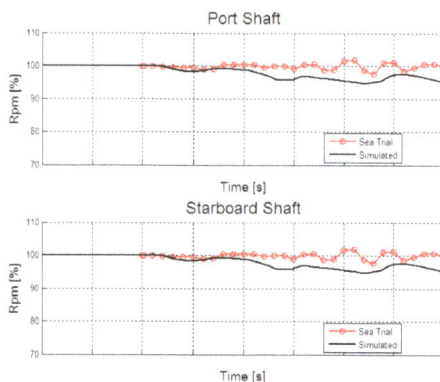

Figure 8.19: Zig Zag – Shaft lines revolution vs. Time

instability. However, in the simulated time history, this behaviour appears from the manoeuvre beginning. The maximum difference between the two trends are below 6%. Similar to the evolution manoeuvre, the difference can be attributed to different control settings, particularly to a propeller pitch reduction function setting.

Fig. 8.20 shows the ship speed time history. Each value is dimensionless with respect to the approaching velocity recorded during the trial. A certain difference between experimental and simulated values is observed, with larger oscillations in the simulation, but in general the two time histories are well-related. It is important to note that, during the sea trial, this channel has been recorded without decimal units, thus providing a discontinuous curve.

Fig. 8.21 shows the transversal heeling over time. Each value (both simulated and experimental) is dimensionless with respect to maximum angle reached during the experimental trial. A good correlation is observed in terms of absolute value and period for the two time histories, proving once again the validity of the manoeuvring

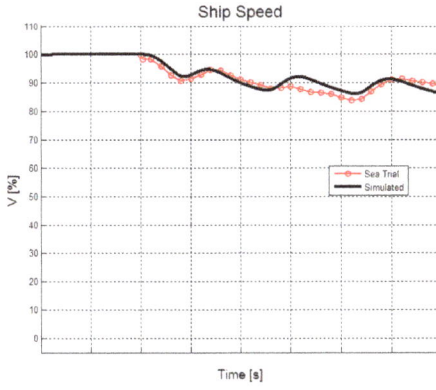

Figure 8.20: Zig Zag – Ship speed vs. Time

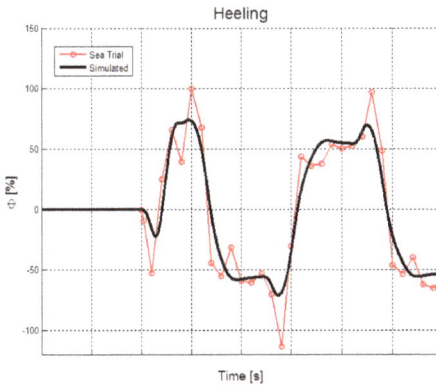

Figure 8.21: Zig Zag – Heeling angle vs. Time

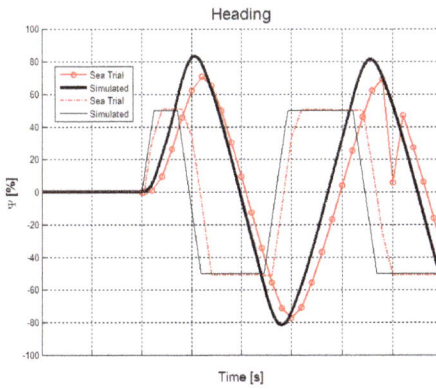

Figure 8.22: Zig Zag – Heading vs. Time

model describing the roll motion described in 5.3 and 5.4. Such as in the evolution manoeuvre, a difference on the peak values still exists.

Fig. 8.22 shows the heading versus time. In Fig. 8.22, the command rudder is also reported in both cases. This helps to evaluate the overshoot angle. The experimental and simulated data are well-related in terms of overshoot angle and oscillation period.

9 Conclusions

The proposed simulation approach, employed at the design stage, gives the designer insight into expected performance of a designed vessel, providing feedback that would otherwise be unavailable until the real system is built and tested.

This insight provides the possibility to check a number of design alternatives to optimize vessel design.

This Ph.D. thesis was inspired from a collaboration between Genova University and an important automation provider. The main target of this work was the development and use of simulation techniques for the propulsion control system design of a naval vessel. Thanks to the previously mentioned collaboration, it has been possible to apply the developed methods and models to a real industrial project. The principal objective has been well-achieved. The use of this kind of simulation platform to test the propulsion controller before the installation onboard enabled the automation provider to significantly reduce tuning time during sea trials. Furthermore, the simulation platform allowed the optimization of propulsion plant performance in terms of responsiveness, fuel consumption, and safety.

After the sea trials, the different models were validated. A good correlation between experimental and simulated data was observed. Obviously, some differences exist, and it is important understand the causes for future improvement. Generally, the differences between experimental and simulated data could be attributed to:
- Different conditions simulated
- Measurement uncertainties
- Missing data
- Assumptions used
- Numerical error

These potential sources for discrepancies could serve as future topics for further investigation in the ship design process.

The methodology and the simulation *"kernel"* developed and validated could become an invaluable multi-purpose tool for the maritime community (design studies, shipyard, automation provider, ship owners, crew, harbor authorities, classification register, etc.) because it allows the following tasks to be completed in a faster, more efficient manner:
- Size and optimize the propulsion plant
- Develop and optimize the propulsion control system
- Improve safety onboard
- Environmental assessment & IMO indicators
- Study maintenance programs
- Crew training
- Avoid collision in narrow water (harbor).

Acknowledgement

First of all, I would like to thank my supervisor, Professor Massimo Figari. His personal investment in me as a scientist when I was completing my MSc thesis and his mentorship led me to attend a Ph.D. course. During the three years I spent to complete Ph.D. thesis, he continuously supported me, which was encouraging especially during hard times. He is a great advisor and, at the same time, a friend.

I would like to extend special thanks also to my office colleague, Prof. Marco Altosole, for his help during the extended time we spent working together.

Additional thanks to Prof. Stefano Vignolo, who gave me the tools to complete the rigorous mathematics required for this work, and to Prof. Michele Viviani for his infinite knowledge of naval architecture topics.

Finally, I would like to express my gratitude to all my family and all my Cumpa's friends for their continuous support and energy. Particular thanks to my parents for their advice.

All this work would have been impossible without someone especially close to me. Thank you, Marta, for your love and patience.

Nomenclature

SYMBOL	UNIT	DESCRIPTION
a	[–]	Propeller inflow factor
$\underline{a}_0$	[m/s^2]	Acceleration of propulsor with respect to $\mathbf{n}$
$\mathbf{b}$	[–]	Ship fixed frame
c	[m]	Rudder average chord
$corr_{CD\_spade}$	[–]	Empirical drag correction for spade rudder
$\mathbf{d}$	[–]	Rudder 0°fixed frame
d_a	[m]	Lever arm of blade frictional axial force
d_{prop_rud}	[–]	Distance between rudder and propeller
dCL	[–]	Rudder lift slope coefficient
$\mathbf{e}$	[–]	Hub fixed frame
$e(t)$	[–]	P.I.D. Error
e_{Osw}	[–]	Oswald correction shape factor
e_{yp}	[m]	Yoke-pin eccentricity
$\mathbf{f}$	[–]	Blade fixed frame
$f(a)$	[–]	Propeller q factor
g	[m/s^2]	Gravity acceleration
h	[J/kg]	Enthalpy
k_a	[–]	Propeller side wash factor
k_{prop}	[–]	Corrective coefficient for propeller fluid velocity
k_s	[–]	Propeller spinner factor
k_{slip}	[–]	Coefficient to reduce propeller slipstream effect
m_0	[–]	Slope of propeller lift curve
m	[kg]	Ship mass
m_b	[kg]	Blade mass
$\dot{m}_i$	[kg/s]	Oil mass flow in
m_m	[kg]	Pitch change mechanism mass
m_o	[kg]	Oil mass
$\dot{m}_o$	[kg/s]	Oil mass flow out
m_p	[kg]	Propeller mass
m_y	[kg]	Yoke mass
$\mathbf{n}$	[kg]	Earth fixed frame
n	[rad/s]	Shaft line rpm
n_{max}	[rpm]	Max shaft line rpm
n_{min}	[rpm]	Min shaft line rpm
n_{set}	[rpm]	Shaft line rpm set-point
$nav\,mode$	[–]	Operative mode
p	[rad/s]	Roll rate
p_1	[Pa]	Astern chamber oil pressure
p_2	[Pa]	Ahead chamber oil pressure
$\dot{p}_{oil}$	[Pa/s]	Time derivative of hub oil pressure
q	[rad/s]	Pitch rate
q_i	[m^3/s]	Oil volumetric flow in
q_{out}	[m^3/s]	Oil volumetric flow out

r	[rad/s]	Yaw rate
s	[m]	Average rudder span
t	[m]	Max rudder thickness
$\tilde{t}$	[–]	Thrust corrective coefficient during manoeuvres
t_0	[m]	Piston head orifice thickness
$threshold$	[rpm]	Crash stop rpm threshold
u	[m/s]	Ship surge velocity
u_{acc}^b	[m/s]	Propeller accelerated flow velocity
u_{prop}	[m/s]	Fluid velocity in propeller disk plane
u_∞	[m/s]	Fluid velocity in upstream region from rudder
$u_{A\infty}$	[m/s]	Fluid velocity at infinity downstream from rudder
v	[m/s]	Ship sway velocity
$\dot{v}_G$	[m/s²]	Ship gravity centre acceleration vector
w	[–]	Rudder fixed frame
w	[m/s]	Ship heave velocity
x	[m]	Longitudinal ship position
x_{CH}	[m]	Longitudinal blade pressure centre position
x_f	[m]	Longitudinal floatation centre position
x_G	[m]	Longitudinal gravity centre position
x_{Max_AH}	[m]	Max stroke ahead
x_{pist}	[m]	Piston stroke
$\dot{x}_{pist}$	[m/s]	Piston velocity
x_{rud}	[m]	Longitudinal rudder position
$\ddot{x}_{pist}$	[m/s²]	Piston acceleration
y	[m]	Transversal ship position
y_{CH}	[m]	Transversal blade pressure centre position
y_G	[m]	Transversal gravity centre position
y_{rud}	[m]	Transversal rudder position
z	[–]	Fluid fixed frame
z	[m]	Vertical ship position
z_{CH}	[m]	Vertical blade pressure centre position
z_G	[m]	Vertical gravity centre position
z_{rud}	[m]	Vertical rudder position
z_{wl}	[m]	Waterline
A_1	[m²]	Yoke area of astern chamber
A_2	[m²]	Yoke area of ahead chamber
A_E	[m²]	Propeller expanded area
A_O	[m²]	Propeller disk area
A_P	[m²]	Rudder area inside the flux tube
A_R	[m²]	Rudder area
A_w	[m²]	Water plane area
AR_{eff}	[–]	Effective rudder aspect ratio
AR_{geom}	[–]	Geometric rudder aspect ratio
B	[Pa]	Oil Bulk modulus
B_p	[Kg/s]	Pitch mechanism damping coefficient
B_s	[N]	Ship buoyancy
B_v	[m]	Vessel breadth

C_A	[–]	Hydrodynamic Coriolis centripetal ship matrix
C_B	[–]	Block coefficient
C_{D0}	[–]	Rudder drag coefficient at zero degree
C_L	[–]	Rudder lift coefficient
C_{ip}	[m^3/(Pa s)]	Leakage coefficient
CPP	[–]	Controllable pitch propellers
C_{RB}	[–]	Coriolis centripetal ship matrix
D	[m]	Propeller diameter
D_L	[–]	Linear damping matrix
D_{NL}	[–]	Non-linear damping matrix
D_P	[m]	Piston head diameter
D_{rud}	[–]	Rudder drag
D_T	[m]	Flow tube diameter
$\underline{F}_{ax}$	[N]	Blade root friction axial force vector
$\underline{F}_{HD}$	[N]	Propulsor hydrodynamic force vector
$\underline{F}_{hyd}$	[N]	Pitch mechanism hydraulic force
$\underline{F}_i$	[N]	External force and moment acting on the vessel
$\underline{F}_I$	[N]	Total inertial forces acting on the propulsor blade
$\underline{F}_{rad}$	[N]	Blade root friction radial force vector
$\underline{F}^C$	[N]	Propulsor Coriolis force vector
$\underline{F}^S$	[N]	Propulsor transportation force vector
$\underline{F}^W$	[N]	Propulsor weight force vector
G	[–]	Restoring matrix
GML	[m]	Longitudinal metacentric height
GMT	[m]	Transversal metacentric height
GZ	[m]	Righting arm
K_D	[–]	Derivative Gain
K_I	[–]	Integral Gain
KMT	[m]	Distance of buoyancy centre from metacentre
K_p	[–]	Hydrodynamic coefficients of D_L matrix
$K_{\dot{p}}$	[–]	Hydrodynamic coefficients of M_A matrix
K_P	[–]	Proportional Gain
$K_{p\|p\|}$	[–]	Hydrodynamic coefficients of D_{NL} matrix
K_Q	[–]	Propulsor torque coefficient
KQ_{SH}	[–]	Hydrodynamic spindle torque coefficient
KQ_{SI}	[–]	Inertial spindle torque coefficient
K_r	[–]	Hydrodynamic coefficients of D_L matrix
$K_{\dot{r}}$	[–]	Hydrodynamic coefficients of M_A matrix
$K_{r\|r\|}$	[–]	Hydrodynamic coefficients of D_{NL} matrix
$K_{r\|r\|\|\varphi\|}$	[–]	Hydrodynamic coefficients of D_{NL} matrix
$K_{r\|v\|}$	[–]	Hydrodynamic coefficients of D_{NL} matrix
$K_{r\|\varphi\|}$	[–]	Hydrodynamic coefficients of D_{NL} matrix
K_T	[–]	Non-dimensional propeller thrust coefficient
K_v	[–]	Hydrodynamic coefficients of D_L matrix
$K_{\dot{v}}$	[–]	Hydrodynamic coefficients of M_A matrix
K_{vrr}	[–]	Hydrodynamic coefficients of D_{NL} matrix
$K_{v\|v\|}$	[–]	Hydrodynamic coefficients of D_{NL} matrix

$K_{v\|v\|\|\varphi\|}$	[–]	Hydrodynamic coefficients of D_{NL} matrix
$K_{v\|\varphi\|}$	[–]	Hydrodynamic coefficients of D_{NL} matrix
I_1	[–]	Propeller side area index
I_b	[kg m²]	Blade Inertia
I_e	[kg m²]	Engine Inertia
I_g	[kg m²]	Gearbox Inertia
I_p	[kg m²]	Propeller Inertia
I_s	[kg m²]	Shaft line polar inertia
J	[–]	Propeller advance coefficient
L	[m]	Ship length
LHV	[J/kg]	Lower heating value
L_{rud}	[–]	Rudder lift
L_p	[m]	Piston head thickness
L_{pp}	[m]	Length between perpendiculars
M	[–]	Inertia and added mass ship matrix
M_A	[–]	Added mass ship matrix
$\underline{M}_O^C$	[Nm]	Blade moments due to Coriolis forces
$\underline{M}_O^I$	[Nm]	Total blade moments due to inertial forces
$\underline{M}_O^S$	[Nm]	Blade moments due to transportation forces
$\underline{M}_O^W$	[Nm]	Blade moments due to weight forces
M_q	[–]	Hydrodynamic coefficients of D_L matrix
$M_{\dot{q}}$	[–]	Hydrodynamic coefficients of M_A matrix
$M_{q\|q\|}$	[–]	Hydrodynamic coefficients of D_{NL} matrix
M_{RB}	[–]	Rigid body inertia ship matrix
N_p	[–]	Hydrodynamic coefficients of D_L matrix
$N_{\dot{p}}$	[–]	Hydrodynamic coefficients of M_A matrix
$N_{p\|p\|}$	[–]	Hydrodynamic coefficients of D_{NL} matrix
N_r	[–]	Hydrodynamic coefficients of D_L matrix
$N_{\dot{r}}$	[–]	Hydrodynamic coefficients of M_A matrix
$N_{r\|r\|}$	[–]	Hydrodynamic coefficients of D_{NL} matrix
$N_{r\|r\|\|\varphi\|}$	[–]	Hydrodynamic coefficients of D_{NL} matrix
$N_{r\|v\|}$	[–]	Hydrodynamic coefficients of D_{NL} matrix
$N_{r\|\varphi\|}$	[–]	Hydrodynamic coefficients of D_{NL} matrix
N_v	[–]	Hydrodynamic coefficients of D_L matrix
N_{vrr}	[–]	Hydrodynamic coefficients of D_{NL} matrix
$N_{v\|v\|}$	[–]	Hydrodynamic coefficients of D_{NL} matrix
$N_{v\|v\|\|\varphi\|}$	[–]	Hydrodynamic coefficients of D_{NL} matrix
$N_{v\|\varphi\|}$	[–]	Hydrodynamic coefficients of D_{NL} matrix
O	[–]	**n** frame origin
O_b	[–]	**b** frame origin
O_d	[–]	**d** frame origin
O_e	[–]	**e** frame origin
O_f	[–]	**f** frame origin
O_w	[–]	**w** frame origin
O_z	[–]	**z** frame origin
OD	[–]	Oil Distribution
P_m	[W]	Engine brake power
$\underline{Q}_{ax}$	[Nm]	Blade root friction axial moment vector

Q_{eng}	[Nm]	Engine torque
Q_{fric}	[Nm]	Frictional shaft torque
Q_{hyd}	[Nm]	Hydraulic torque
Q_i	[Nm]	External torques acting on the shaft
Q_p	[Nm]	Propeller torque
Q_{port}	[Nm]	Port shaft torque
$\underline{Q}_{rad}$	[Nm]	Blade root friction radial moment vector
Q_S	[Nm]	Total spindle torque
Q_{SF}	[Nm]	Frictional component of spindle torque
Q_{SH}	[Nm]	Hydrodynamic component of spindle torque
Q_{SI}	[Nm]	Inertial component of spindle torque
Q_{stbd}	[Nm]	Starboard shaft torque
$Q_{-\Phi}$	[Nm]	Interaction torque between blade and its bearing
R_b^n	[–]	Rotational Matrix from **b** to **n** (linear velocity)
R_b^d	[–]	Rotational Matrix from **b** to **d** (linear velocity)
R_d^w	[–]	Rotational Matrix from **d** to **w** (linear velocity)
R_w^z	[–]	Rotational Matrix from **w** to **z** (linear velocity)
S	[–]	Cross-product operator
T	[N]	Propulsor thrust
T_S	[m]	Ship draft
T_b^n	[–]	Rotational Matrix from **b** to **n** (angular velocity)
Tel	[–]	Step lever position
V	[m/s]	Ship speed
V_a	[m/s]	Propulsor advance velocity
V_c	[m^3]	Actuating chamber volume
V_0	[m^3]	Initial actuating chamber volume
W_s	[N]	Ship weight force
X_{rr}	[–]	Hydrodynamic coefficients of D_{NL} matrix
X_u	[–]	Hydrodynamic coefficients of D_L matrix
$X_{\dot{u}}$	[–]	Hydrodynamic coefficients of M_A matrix
X_{vr}	[–]	Hydrodynamic coefficients of D_{NL} matrix
X_{vv}	[–]	Hydrodynamic coefficients of D_{NL} matrix
Y_p	[–]	Hydrodynamic coefficients of D_L matrix
$Y_{\dot{p}}$	[–]	Hydrodynamic coefficients of M_A matrix
$Y_{p|p|}$	[–]	Hydrodynamic coefficients of D_{NL} matrix
Y_r	[–]	Hydrodynamic coefficients of D_L matrix
$Y_{r|r|}$	[–]	Hydrodynamic coefficients of D_{NL} matrix
$Y_{r|r||\varphi|}$	[–]	Hydrodynamic coefficients of D_{NL} matrix
$Y_{r|v|}$	[–]	Hydrodynamic coefficients of D_{NL} matrix
$Y_{r|\varphi|}$	[–]	Hydrodynamic coefficients of D_{NL} matrix
Y_v	[–]	Hydrodynamic coefficients of D_L matrix
$Y_{\dot{v}}$	[–]	Hydrodynamic coefficients of M_A matrix
Y_{vrr}	[–]	Hydrodynamic coefficients of D_{NL} matrix
$Y_{v|v|}$	[–]	Hydrodynamic coefficients of D_{NL} matrix
$Y_{v|v||\varphi|}$	[–]	Hydrodynamic coefficients of D_{NL} matrix
$Y_{v|\varphi|}$	[–]	Hydrodynamic coefficients of D_{NL} matrix
Z	[–]	Number of blades
ZB	[m]	Distance between buoyancy centre and keel line

ZG	[m]	Distance between gravity centre and keel line
Z_w	[–]	Hydrodynamic coefficients of D_L matrix
$Z_{\dot{w}}$	[–]	Hydrodynamic coefficients of M_A matrix
$Z_{w\lvert w\rvert}$	[–]	Hydrodynamic coefficients of D_{NL} matrix
α_{PEAK}	[rad]	Rudder max lift angle
α_{STALL}	[rad]	Rudder stall angle
β	[rad]	Drift angle
β_{loc}	[rad]	Drift angle at rudder
y	[–]	Fluid straightening vector
δ	[rad]	Rudder angle
δ_{eff}	[rad]	Rudder effective attack angle
δ_{fix}	[rad]	Rudder axis fixed inclination
η	[m - rad]	Ship position vector
$\dot{\eta}$	[m/s - rad/s]	Ship velocity vector in **n** -basis
$\ddot{\eta}$	[m/s^2 - rad/s^2]	Ship acceleration vector in **n** -basis
θ	[rad]	Pitch angle
κ_f	[–]	Frictional coefficient of the blade-hub
μ	[Pa s]	Oil dynamic viscosity
v	[m/s - rad/s]	Ship velocity vector in **b** -basis
$\underline{v}_G$	[m/s]	Linear velocity vector for the propulsor gravity centre
$\underline{v}_{rud}$	[m/s]	Fluid velocity vector in **b** -basis on rudder
$\dot{v}$	[m/s^2 - rad/s^2]	Ship acceleration vector in **b** -basis
$\underline{v}^b_{b/n}$	[m/s]	Ship linear velocity vector
ξ	[–]	Whole set of points constituting the blade
ρ	[kg/m^3]	Sea water mass density
ρ_b	[kg/m^3]	Blade mass density
ρ_{oil}	[kg/m^3]	Oil mass density
σ	[–]	Blade solidity
τ	[N - Nm]	External force and moment acting on the vessel
τ_A	[N - Nm]	Added mass forces and moments vector
τ_D	[N - Nm]	Damping forces and moments vector
τ_G	[N - Nm]	Restoring forces and moments vector
τ_H	[N - Nm]	Hydrodynamic hull forces and moments vector
τ_L	[N - Nm]	Linear damping forces and moments vector
τ_{NL}	[N - Nm]	Non-linear damping forces and moments vector
τ_P	[N - Nm]	Propulsor hydrodynamic forces and moments vector
τ_{PF}	[N - Nm]	Propulsor frictional forces and moments vector
τ_{PI}	[N - Nm]	Propulsor inertial forces and moments vector
τ_{PROP}	[N - Nm]	Total forces and moments acting on the propulsors
τ_R	[N - Nm]	Rudders forces and moments vector
τ_W	[N - Nm]	Environmental forces and moments vector
φ	[rad]	Propeller pitch angle
$\dot{\varphi}$	[rad/s]	Blade angular velocity
$\ddot{\varphi}$	[rad/s^2]	Blade angular acceleration
φ_0	[rad]	Zero thrust pitch angle
φ_{Max_AH}	[rad]	Max propeller pitch ahead
ϕ	[rad]	Roll angle

χ	[rad]	Course angle
ψ	[rad]	Yaw angle
$\underline{\omega}_{b/n}^{b}$	[rad/s]	Ship angular velocity vector
$\underline{\omega}_B$	[rad/s]	Blade angular velocity vector
$\underline{\omega}_H$	[rad/s]	Hub angular velocity vector
$\dot{\omega}_s$	[rad/s^2]	Shaft line acceleration
Δ_r	[m]	Propeller flow radius increment
Δ_{rpm}	[rpm]	Slam Start tolerance
Δ_w	[–]	Asymmetrical wake field increment
Φ	[N]	Interaction force between blade and its bearing
∇	[m^3]	Displaced volume
$(1 - t)$	[–]	Thrust deduction factor
$(1 - w)$	[–]	Wake factor
$(CH - O)$	[m]	Position vector of the blade pressure centre
$(G - O)$	[m]	Position vector of the blade gravity centre
$(P - O)$	[m]	Generic distance vector

Appendix

A

$$\underline{M}_O^C = -2 \int_\beta \bigwedge \rho \, (P - O) \bigwedge \left[\underline{\omega}_H \bigwedge \left[\underline{\omega}_\varphi \bigwedge (P - O) \right] \right] d\tau \qquad \text{(A.1)}$$

$\underline{M}_O^C$ is the moment (with respect to the point O) of the Coriolis force acting on a single blade. Using the skew-symmetric property of the vector (wedge) product between vectors ($\underline{a} \bigwedge \underline{b} = -\underline{b} \bigwedge \underline{a}$) as well as the well-known formula $\underline{a} \bigwedge (\underline{b} \bigwedge \underline{c}) = \underline{b} \, (\underline{a} \cdot \underline{c}) - \underline{c} \, (\underline{a} \cdot \underline{b})$ (the dot denoting scalar product between vectors), we can work out the vector expression under the integral A.1 in different ways. In particular, we obtain the identities:

$$(P - O) \bigwedge \left[\underline{\omega}_H \bigwedge \left[\underline{\omega}_\varphi \bigwedge (P - O) \right] \right] = $$
$$\underline{\omega}_\varphi \bigwedge \left[(P - O) \bigwedge \left[\underline{\omega}_H \bigwedge (P - O) \right] \right] - \underline{\omega}_\varphi \bigwedge \underline{\omega}_H \, |P - O|^2 \qquad \text{(A.2)}$$

and

$$(P - O) \bigwedge \left[\underline{\omega}_H \bigwedge \left[\underline{\omega}_\varphi \bigwedge (P - O) \right] \right] = -\underline{\omega}_\varphi \bigwedge (P - O) \left[(P - O) \cdot \underline{\omega}_H \right] \qquad \text{(A.3)}$$

Moreover, it is easily seen that the following identities hold as well:

$$\underline{\omega}_\varphi \bigwedge (P - O) \left[(P - O) \cdot \underline{\omega}_H \right] = $$
$$(P - O) \bigwedge \left[\left(\underline{\omega}_\varphi \bigwedge \underline{\omega}_H \right) \bigwedge (P - O) \right] + \underline{\omega}_H \bigwedge (P - O) \left[\underline{\omega}_\varphi \cdot (P - O) \right] \qquad \text{(A.4)}$$

and

$$\underline{\omega}_H \bigwedge \left[(P - O) \bigwedge \left[\underline{\omega}_\varphi \bigwedge (P - O) \right] \right] = \underline{\omega}_H \bigwedge \underline{\omega}_\varphi \, |P - O|^2 - \underline{\omega}_H \bigwedge (P - O) \left[\underline{\omega}_\varphi \cdot (P - O) \right] \qquad \text{(A.5)}$$

Combining A.2, A.3, A.4, and A.5, we then obtain:

$$-2 \, (P - O) \bigwedge \left[\underline{\omega}_H \bigwedge \left[\underline{\omega}_\varphi \bigwedge (P - O) \right] \right] = $$
$$-\underline{\omega}_\varphi \bigwedge \left[(P - O) \bigwedge \left[\underline{\omega}_H \bigwedge (P - O) \right] \right] + \underline{\omega}_\varphi \bigwedge \underline{\omega}_H \, |P - O|^2 + $$
$$+ (P - O) \bigwedge \left[\left(\underline{\omega}_\varphi \bigwedge \underline{\omega}_H \right) \bigwedge (P - O) \right] + \underline{\omega}_H \bigwedge (P - O) \left[\underline{\omega}_\varphi \cdot (P - O) \right] = $$
$$-\underline{\omega}_\varphi \bigwedge \left[(P - O) \bigwedge \left[\underline{\omega}_H \bigwedge (P - O) \right] \right] + (P - O) \bigwedge \left[\left(\underline{\omega}_\varphi \bigwedge \underline{\omega}_H \right) \bigwedge (P - O) \right] $$
$$-\underline{\omega}_H \bigwedge \left[(P - O) \bigwedge \left[\underline{\omega}_\varphi \bigwedge (P - O) \right] \right] \qquad \text{(A.6)}$$

Inserting the content of Eq. A.6 into Eq. A.1 and recalling the general definition of the action of the inertia tensor on a vector $\underline{a}$, we obtain:

$$I_O(\underline{a}) = \int_\beta \rho\,(P-O) \bigwedge \left[\underline{a} \bigwedge (P-O)\right] d\tau \qquad (A.7)$$

Then, taking into account that the vectors $\underline{\omega}_P$ and $\underline{\omega}_\varphi$ are independent of the integration variables, we reach the final expression:

$$\underline{M}_O^C = -\underline{\omega}_\varphi \bigwedge I_O(\underline{\omega}_H) - \underline{\omega}_H \bigwedge I_O\left(\underline{\omega}_\varphi\right) + I_O\left(\underline{\omega}_\varphi \bigwedge \underline{\omega}_H\right) \qquad (A.8)$$

where I_O denotes the inertia tensor of the single blade with respect to the point O. Concerning the moment of the transportation force with respect to the point O, we have:

$$\underline{M}_O^S = -\int_\beta \rho\,(P-O) \bigwedge \left[\underline{a}_O + \underline{\omega}_H \bigwedge \left[\underline{\omega}_H \bigwedge (P-O)\right] + \underline{\dot{\omega}}_H \bigwedge (P-O)\right] d\tau \qquad (A.9)$$

On the one hand, it is immediately seen that:

$$-\int_\beta \rho(P-O) \bigwedge \underline{a}_O\, d\tau = -M\,(G-O) \bigwedge \underline{a}_O \qquad (A.10)$$

and

$$-\int_\beta \rho\,(P-O) \bigwedge \left[\underline{\dot{\omega}}_H \bigwedge (P-O)\right] d\tau = -I_O(\underline{\dot{\omega}}_H) \qquad (A.11)$$

On the other hand, we have the identity:

$$(P-O) \bigwedge \left[\underline{\omega}_H \bigwedge \left[\underline{\omega}_H \bigwedge (P-O)\right]\right] = (P-O) \bigwedge [\underline{\omega}_H\,(\underline{\omega}_H \cdot (P-O))] =$$
$$(P-O) \bigwedge [\underline{\omega}_H\,(\underline{\omega}_H \cdot (P-O))] + \underline{\omega}_H \bigwedge \underline{\omega}_H\,|P-O|^2 =$$
$$\underline{\omega}_H \bigwedge \left[-(P-O)\,(\underline{\omega}_H \cdot (P-O)) + \underline{\omega}_H\,|P-O|^2\right] = \underline{\omega}_H \bigwedge \left[(P-O) \bigwedge \left[\underline{\omega}_H \bigwedge (P-O)\right]\right] \qquad (A.12)$$

From Eq. A.12, it follows that:

$$-\int_\beta \rho\,(P-O) \bigwedge \left[\underline{\omega}_H \bigwedge \left[\underline{\omega}_H \bigwedge (P-O)\right]\right] d\tau = -\underline{\omega}_H \bigwedge I_O(\underline{\omega}_H) \qquad (A.13)$$

Taking Eqs. A.10, A.11, and A.12 into account, we can rewrite Eq. A.9 in the final form:

$$\underline{M}_O^S = -M\,(G-O) \bigwedge \underline{a}_O - \underline{\omega}_H \bigwedge I_O(\underline{\omega}_H) - I_O(\underline{\dot{\omega}}_H) \qquad (A.14)$$

Bibliography

[1]   Altosole M., Benvenuto G., Campora U.(2012), "Numerical modelling of the engines governors of a CODLAG propulsion plant", Proceedings of the 10th International Conference on Marine Sciences and Technologies, Black Sea 2010, Varna, Bulgaria, 7-9 October, 2010. Pages 173-178. ISSN: 1314-0957.

[2]   Altosole M., Benvenuto G., Campora U., Figari M., (2009), "Real-Time Simulation of a COGAG naval Ship Propulsion System" - Journal of Engineering for the Maritime Environment, Vol. 223, N. 1.

[3]   Altosole M., Benvenuto G., Campora U., Figari M., Bagnasco A., D'Arco S., Giuliano M., Giuffra V., Spadoni A., Zanichelli A., Michetti S., Ratto M., (2008), "Real Time simulation of the propulsion plant dynamic behaviour of the Aircraft Carrier Cavour", INEC 2008, Hamburg.

[4]   Altosole M., Benvenuto G., Galli M., Martelli M.,(2011a), "Advances in automation design for fast vessels propulsion", Proceedings - 9th Symposium on High Speed Marine Vehicles, HSMV 2011, Naples, Italy, May 25-27, 2011. ISBN: 978889061120.

[5]   Altosole M., Boote D., Brizzolara S., Viviani M. (2013): "Integration of Numerical Modelling and Simulation Techniques for the Analysis of Towing Operations of Cargo Ships", International Review of Mechanical Engineering (IREME), Volume 7, Number 7, November 2013. Print ISSN: 1970 – 8734.

[6]   Altosole M., Dubbioso G., Figari M., Viviani M., Michetti S., Millerani Trapani A., (2010b), "Simulation of the dynamic behaviour of a CODLAG propulsion plant", Warship 2010 Conference, London, UK.

[7]   Altosole M., Benvenuto G., Figari M.(2005): "Performance Prediction of a Planning Craft by Dynamic Numerical Simulation", Proceedings – 7th Symposium on High Speed Marine Vehicles, HSMV 2005, Naples, Italy, September 21-23, 2005. ISBN 88-88987-02-9.

[8]   Altosole M., Figari M.,(2011b), "Effective simple methods for numerical modelling of marine engines in ship propulsion control systems design", Journal of Naval Architecture and Marine Engineering, December 2011, pages 129-147. DOI 10.3329/ jname.v8i2.7366.

[9]   Martelli M., Figari M., Altosole M., Vignolo S.(2013): "Controllable pitch propeller actuating mechanism, modelling and simulation", Proceedings of the Institution of Mechanical Engineers, Part M: Journal of Engineering for the Maritime Environment, DOI: 10.1177/1475090212468254, published online 11 January 2013, pages 1-15..

[10]   Altosole M., Figari M., Marcilli G., Michetti S., Ratto M., Spadoni A.: "Ship control system wide integration and the use of dynamic simulation techniques in the Fremm project", Proceedings of the International Conference on Electrical Systems for Aircraft, Railway and Ship propulsion, ESARS'10, Bologna, Italy, October 19-21, 2010. ISBN: 978-1-4244-9094-3.

[11]   Ankudinov, V., Kaplan, P., Jacobsen, B., (1993), "Assessment and Principal Structure of the Modular Mathematical Model for Ship Manoeuvrability Predictions and Real-Time Manoeuvring Simulation", Proceeding of MARSIM 93, pp. 40-51, St. Johns, Canada.

[12]   Bakker J.C., Grimmelius H.T., Wesselink A.F.,(2006), "The use of non linear models in the analysis of CPP actuator behaviour", ICMES Conference.

[13]   Benvenuto G., Brizzolara S., Figari M., (2001), "Simulation Of The Propulsion System Behaviour During Ship Standard Manoeuvres", PRADS 2001.

[14]   Benvenuto G., Campora U. (2005a), "A Gas Turbine Modular Model for Ship Propulsion Studies" - HSMV, 7th Symposium on High Speed Marine Vehicles, Naples, Italy, 21 - 23 September, 2005.

[15]   Benvenuto G., Campora U., Carrera G., Figari M., (2005b), "Interaction Between Ship Propulsion Plant Automation and Simulation", IMAM 2005 Conference Proceedings, Vol. 1.

[16] Bodnaruk A. & Rubis C.J., (1971), "Acceleration Performance Analysis of a Gas Turbine De-stroyer Escort", Journal of Engineering for Power.

[17] Bossel, H., (1994), "Modelling and Simulation", book edit by A.K. Peters, Wellesley (USA), ISBN: 1-56881-033-4.

[18] Brix, J., (1993), "Manoeuvring Technical Manual", book edit by Seehafen Verlag GmbH, Ham-burg, ISBN 3-87743-902-0.

[19] Carlton J., (2007), "Marine propellers and propulsion", ISBN 978-07506-8150-6.

[20] Chu C., Chan Z.L., She Y.S., Yuan V.Z., (1979), "The 3-bladed JD-CPP series", 4th Lips Propeller Symposium", p. 53-79, Drunen.

[21] Coraddu, A., Dubbioso, G., Guadalupi, D., Mauro, S., Viviani, M., (2012), "Experimental in-vestigation of asymmetrical propeller behaviour of twin screw ships during manoeuvres", MARSIM2012, Singapore 23-27 April 2012.

[22] Figari M., Guedes Soares C., (2009), "Fuel consumption and exhaust emissions reduction by dynamic pitch control", Analysis and Design of Marine Structures. Lisbon, Portugal, March 2009, London: Taylor & Francis, p. 543-550, ISBN/ISSN: 978-0-415-54934-9.

[23] Fossen T.I., (1994), "Guidance and Control of Ocean Vehicles", John Wiley and sons Ltd. ISBN 0-471-94113-1

[24] Fossen T.I., (2002), "Marine Control System", Marine Cybernetics, ISBN/82-92356-00-2

[25] Fossen T.I.,(2011), "Handbook of marine craft hydrodynamics and motion control", Editor Wi-ley, ISBN:9781119991496.

[26] Fujii, H., Tsuda, T., (1961), "Experimental Researches on Rudder Performance" (in Japanese), Society of Naval Architects of Japan, Vol.110, pp 31-45.

[27] Gaggero S., Brizzolara S., (2009), "A panel method for transcavitating marine propellers",7th International symposium on cavitation (CAV2009), Ann Arbor, MI, August 2009.

[28] Gaggero S., Villa D., Brizzolara S.,(2010), "RANS and panel method for unsteady flow propeller analysis", Journal of Hydrodynamic; Vol. 22, pp. 564-569.

[29] Godjevac, M., Van Beek, T., Grimmelius, H.T., Tinga, T. & Stapersma, D., (2009), "Prediction of fretting motion in a controllable pitch propeller during service", Journal of Engineering for the Maritime Environment, Volume 223.

[30] Kijima, K., Furukawa, Y., (1998), "Effect of roll motion on manoeuvrability of ship", International symposium and workshop on force acting on a manoeuvring vessel, pp. 157-163, September 1998, Val de Reuil, France.

[31] Molland, A.F. and Turnock, S.R. (1994) "Developments in modelling ship rudder-propeller inter-action", in, Brebbia, C.A., Murthy, T.K.S, Wilson, P.A. and Washams, P. (eds.) Marine, Offshore and Ice Technology. Ashurst, UK, WIT Press, 255-263. (Transactions of the Wessex Institute, 5).

[32] Molland, A.F. and Turnock, S.R., (2007), "Marine rudders and control surfaces: principles, data, design and applications", Oxford, UK, Butterworth-Heinemann, pp. 386.

[33] Pivano, L., (2008), "Thrust Estimation and Control of Marine Propellers in Four-Quadrant Oper-ations", PhD-thesis, Faculty of Engineering Science & Technology, ISBN 978-82-471-6261-3.

[34] Principle of Naval Architecture (PNA) volume 2, Society of Naval Architects and Marine Engi-neers (U.S.), Editors Henry Eastin Rossell, L. B. Chapman.

[35] Rana R. K., Damaodaran K. A., Kang H. S., Gokhale S.S., (1996), "Governing of ship propulsion gas turbine engine in a Seaway", Eleventh Ship Control Systems Symposium, Southampton.

[36] Ribner, H., (1943), "Notes on propeller and slipstream", Langley Aeronautic Memorial Labora-tory, Report n°L4Il2a.

[37] Rubis C.J., (1978). "Guide for Centralized Control and Automation of Ship's Gas Turbine Propul-sion Plant", SNAME T&R Bulletin 3-29.

[38] Rubis C.J., Harper T.R.,(1986), "Governing Ship Propulsion Gas Turbine Engines", SNAME Transactions, Vol. 94.

[39] Sagatun, S. U., Fossen T.I.,(1991), "Lagrangian formulation of underwater vehicles dynamics" Proc of IEEE conference on system, man and cybernetics, Charlottesville, VA, pag 1029-1034.

[40] Smogeli, N., O., (2006), "Control of Marine Propellers From Normal to Extreme Conditions", PhD-thesis, Faculty of Engineering Science & Technology, ISBN 82-471-8147-9.

[41] SNAME (1950), The society of naval architectures and marine engineers. "Nomenclature for treating the motion in submerged body trough a fluid". In: Technical and research bulletin N°1-5.

[42] Viviani .M, Podenzana Bonvino C., Mauro S., (2007), "Analysis of asymmetrical shaft power increase during tight manoeuvre",10th International symposium on practical design of ships and other floating structures (PRADS), Houston, TX, October 2007.

[43] Viviani M, Dubbioso G, Soave M, Notaro C, DePascale R.,(2009) "Hydrodynamic coefficients regressions analysis and application to twin screw vessels", 13th Congress of Intl. Maritime Assoc. of Mediterranean (IMAM 2009), Istanbul, Turkey, 2009.

[44] Wessenlink A.F., Stapersma D., van de Bosch D., Teerhuis P.C., (2006), "Non linear aspect of propeller pitch control", ICMES Conference.

[45] Whicker, L.F., Fehlner, L.F. ,(1958), "Free Stream characteristics of a family of low aspect ratio, all movable control surfaces for application to ship design", David Taylor Model Basin, Report 933.

[46] Wind J.,(1978), "Principles of mechanism used in controllable pitch propellers", 3th Lips Propeller Symposium, pp. 53-66, Drunen.

Index

www.ingramcontent.com/pod-product-compliance
Lightning Source LLC
Chambersburg PA
CBHW040121120426
42814CB00009B/339